AUDIO VISUAL WORLD

Cinema, Theatre & Concert Hall

视听世界：影剧院与音乐厅

高迪国际出版有限公司 编
刘巧妹 郝诗婷 肖新瑶 解静 吴晓瓅 汤雨倩 译

大连理工大学出版社
Dalian University of Technology Press

图书在版编目(CIP)数据

视听世界：影剧院与音乐厅 / 高迪国际出版有限公司编；刘巧妹等译. 一大连：大连理工大学出版社，2012.7

ISBN 978-7-5611-6991-9

Ⅰ.①视… Ⅱ.①高… ②刘… Ⅲ.①电影院－建筑设计－世界②剧院－建筑设计－世界③音乐厅－建筑设计－世界 Ⅳ.①TU242.2

中国版本图书馆CIP数据核字（2012）第122334号

出版发行：大连理工大学出版社
　　　　　（地址：大连市软件园路80号　邮编：116023）
印　　刷：利丰雅高印刷（深圳）有限公司
幅面尺寸：245mm×320mm
印　张：23
插　页：4
出版时间：2012年7月第1版
印刷时间：2012年7月第1次印刷
责任编辑：刘　蓉
责任校对：李　雪
封面设计：屈舒丽

ISBN 978-7-5611-6991-9
定　　价：328.00元

电　话：0411-84708842
传　真：0411-84701466
邮　购：0411-84703636
E-mail: designbooks_dutp@yahoo.cn
URL: http://www.dutp.cn

如有质量问题请联系出版中心：（0411）84709246　84709043

Freight & Salvage Opera House Oslo Opera House Oslo
Amphion Theatre Amphion Theatre
Sauveniere Cinema Corby Cube
Multikino Stary Browar Marlowe Theatre
Kronverk Cinema South Miami-dade Cultural Center
New Theatre in Montalto Di Castro
Opera House Oslo Burlington Performing Arts Centre
Amphion Theatre Music Hall Eindhoven
Corby Cube Freight & Salvage Freight
Marlowe Theatre Sauveniere Cinema
South Miami-dade Cultural Center Inox Multikino Stary Browar Sauv
New Theatre in Montalto Di Castro Kronverk Cinema Inox
Burlington Performing Arts Centre Opera House Oslo Kron
Music Hall Eindhoven Amphion Theatre Ope
Marlowe Theatre Corby Cube Amph
Freight & Salvage Opera House Oslo South Miami-dade Cultural Center
Sauveniere Cinema Amphion Theatre New Theatre in Montalto Di Castro
Multikino Stary Browar Marlowe Theatre Burlington Performing Arts Centre
Kronverk Cinema South Miami-dade Cultural Center Music Hall Eindhoven
Opera House Oslo New Theatre in Montalto Di Castro Opera House Oslo Marlowe T
Amphion Theatre Burlington Performing Arts Centre Amphion Theatre South Mi
Corby Cube Freight & Salvage
Miami-dade Cultural Center Sauveniere Cinema
New Theatre in Montalto Di Castro Inox Multikino Stary Browar Burling
ington Performing Arts Centre
Music Hall Eindhoven Kronverk Cinema Mus
ght & Salvage Freight &
Sauveniere Cinema Opera House Oslo Sauven
Multikino Stary Browar Amphion Theatre Inox
Kronverk Cinema Corby Cube Kronver
Opera House Oslo Marlowe Theatre Opera
Amphion Theatre South Miami-dade Cultural Center Amphion
Corby Cube New Theatre in Montalto Di Castro Marlowe Theatre
Miami-dade Cultural Center South Mi
New Theatre in Montalto Di Castro
ington Performing Arts Centre New Theatre in Montalto Di Castro

PREFACE_1
序言—1

Mr. Eranna Yekbote, Chief Architect,
Era Architects, Mumbai, India

"There is no business like show business" goes a famous jargon, and how true is that, since generations together, movies have enthralled billions of audiences across the world and still are considered the best mode of entertainment as well as an art form often reflecting societal ethos and culture.

Preceding films, entertainment in all cultures was in form of plays, dramas, which gave birth to theatres, housing large audiences. The earliest being the Greeks, followed by the Romans who built theatres on a grand scale which are Architectural marvels till date.

This saw an emergence of theatre architecture which has ever since undergone transformation directly related to the new inventions and developing technology, the most important being the invention of motion picture. It began with stringing 2-dimensional objects moving on screen, to silent black and white movies furthering to sound and eventually colored wide screen movies, the latest being 3D movies.

Modern movie theatres were still one screen till the concept of multiplexes began rapidly transforming the entertainment industry.

Considering the space constraints in busy cities, having more number of screens and seating maximum number of people at the same time watching different movies has become convenient as well as is a lucrative market.

Today a multiplex is an environ people go to not only to watch movies but to get complete entertainment, which has in turn, seen the emergence of food courts, gaming zones, turning it into a social hub.

Hence multiplex design has now become a design challenge where the designer has to take care of private and public spaces, user circulation and safety, keeping in mind the design aesthetic and commercial viability of spaces.

There is a public and private element to cinema design which needs to have clearly defined boundaries of free and restricted movement respectively.

Planning multiple screens on the same floor makes user circulation of prime importance as well as a challenge for structural design. The screens can range from minimum 2 to 3 to a maximum of 8 to even more, which makes it exciting for the designer to explore various design permutations. Especially planning of seats inside the auditorium, acoustics, and the overall look and ambience keeping in mind the user comfort is challenging.

The public areas like cinema entrance lobby, entertainment zone allow the designer to have a free hand at design where colors, forms, materials can be explored blending it with light effects creating an inviting and entertaining ambience.

This book is a compilation of such varied cinema projects ranging from multiple screens to stand-alone cinemas giving a glimpse of modern design aesthetic, which is truly an enjoyable reading experience. Hope you enjoy it too.

Text by vishalakshi M. Subedar

就像那句有名的行话说的一样，"没有比娱乐业更好的行业了"。的确，几代以来，电影已经吸引了全世界亿万的观众，而且至今仍被看做是最好的娱乐方式，同时也是反映社会风气和文化的一种艺术形式。

在电影出现之前，各国的娱乐形式都是戏剧，与此同时，能容纳大量观众的剧院应运而生。剧院最早出现于希腊，之后罗马人开始大规模兴建剧院，其中一些建筑至今仍旧令人叹为观止。

这一时期，剧院建筑兴起。从此，它的发展变化一直与新发明和胶卷的冲印技术革新息息相关，尤为重要的是电影的发明。它开始于屏幕上的2D动态画面，到黑白无声电影，再到有声的宽屏彩色画面，最后是现在的3D电影。

现代电影院仍然是单屏，直到多功能影院理念在娱乐业迅速掀起巨浪。

考虑到喧嚣城市的有限空间，通过提供更多的屏幕和座位，让观众在同一时间观看不同的电影，既方便观众，又是生财之道。

如今，人们去多功能影院不但可以看电影，而且可以得到完全的娱乐和休闲。反过来，这也催生了小吃店、游戏区，使得电影院变成了社交中心。

因此，多功能设计变成了一种设计的挑战，设计师不但需要考虑到私人空间和公共空间、用户循环和安全保障，而且要设计出集美观和商业性于一体的场所。

影院设计要考虑公共和私人要素，而且需要清晰地界定出自由空间和限制空间各自的范围。

在同一层设计多个屏幕需要考虑到用户循环的重要性，同时这也是对结构设计的挑战。屏幕可从最少的两到三个，到最多的八个，甚至更多，这让设计师们感受到发掘多种设计模式的兴奋。尤其是礼堂座位的设计、声效和整个的布景及氛围，都要满足观影者的舒适需求，这对设计师也是个挑战。

像影院入口走廊和娱乐区这样的公共区域，设计师可以不受约束地大胆设计，尝试颜色、布局、材料和灯光的混合使用，以便营造出富有吸引力、令人愉快的氛围。

本书汇集了各种各样的影院建筑设计项目，无论是多屏幕影院还是独立影院，都可以让您对现代美学设计有个大致的了解，这将是个令人愉快的阅读过程。希望您也能喜欢。

PREFACE_2
序言—2

Keith R Williams
Founder and Director of Design
Keith Williams Architects, London

我很荣幸能有机会在职业生涯中设计了大量的文化建筑。我不得不仔细考虑文化本身的含义，以求将建筑概念化，实现通过建筑形式表达文化内涵的目的。我还发现深入了解文化和艺术的历史发展很重要，这样才能以建筑手法去回应其内涵。

本书通过考察大量的重要文化建筑，以及敢于承接这些极具挑战性设计项目的建筑设计师的设计手法，来重点介绍表演艺术和影院建筑。

首先，我想对这两个存在根本差异的艺术形式——影院和剧院，进行区分。两者都是容纳观众的公共演出场所（尽管与剧院不同，电影院也可以在家中体验）。

相比之下，影院的历史较短。自19世纪末期影院出现以来，它就在电视的产生（20世纪中期）和近代的数字化时代的背景下重新定义了自己。然而，去影院观赏却基本保持着本质上的不变性和被动性。观众观看通过中间媒介投射到银幕上的内容——也就是电影。其实并没有真人在放映厅里表演，播放的内容不过是人们在别处进行的陈述或者上演的事件。这种形式虽然很有感染力和吸引力，但是却没法让观众参与其中。它需要的只是简单的场所、优质的声效和良好的放映系统，以此来达到令人满意的播放效果。

剧院整体要更加复杂。戏剧是一种存在了几千年的古老的艺术表现形式。其基本原理就是观众参与的本质，观众与演员之间进行的互动。这种共同经历式的体验与被动的影院截然不同。影院和剧院对建筑上的限制也是不同的。

伟大的戏剧导演Peter Brook凝练出戏剧的本质："一个空荡荡的屋子，我称它为空旷的舞台。一个人横穿这间空屋子，而其他的人看着他。这就是从事戏剧表演所需要的一切。"

然而我们的这一传统是源于古希腊戏剧、西方歌剧或中国歌剧，这些艺术形式需要大批观众、精美布景的舞台、复杂的观众席，以及大面积综合性的后台和技术。这样的复杂性要求建筑师能够独具匠心地进行创造性设计。表演艺术被视为文化成就的巅峰之一，因此，宏伟的文化建筑通常会受到很高的赞誉，能与政府大楼及都市中备受瞩目的朝拜场所媲美。城市规划者和政客则利用这些建筑的独特性作为他们城市的标志。

本书所展示的是近期重要的文化建筑设计作品，也是建筑师和客户挑战文化建筑设计的明证。

I am fortunate enough to have been asked to design a considerable number of cultural buildings during my career. I have had to consider very carefully what is meant by culture itself in order to be able to conceptualise and realise the buildings in which culture can take place. I have also found it important to have a profound sense of the historical path on which culture and the Arts have travelled in order to be able to respond architecturally to its context.

This book considers, through examination of a number of important buildings ways in which architects have responded to the challenge of designing important cultural buildings, specifically those for the Performing Arts and for Cinema.

I want to draw distinction between these two fundamentally different art forms, Cinema and Theatre. Both are usually enacted in public spaces both with an audience, (though cinema unlike theatre can be experienced at home).

Cinema is comparatively recent. From its origins in the late 19th century, it has redefined itself in the context of the arrival of the television (mid 20th century) and more recently the digital age. The experience, however, of watching cinema remains essentially unchanged and passive. The audience reacts to what is projected onto a screen via an interlocutory medium – the film. There are no real people performing in the room, merely a representation of people acting and events staged, elsewhere. It can be powerful and engaging but it is not participatory. It needs only the simplest of spaces, and a good sound and projection system for it to be satisfactorily performed.

Theatre is altogether more complex. It is an ancient art form several thousand years old. Fundamental is its participatory nature, an engagement between actor and audience, a shared experience as distinct from the passivity of the cinema. The architectural constraints are also very different.

The great theatre director Peter Brook distilled the essence of theatre thus "I can take an empty space and call it a bare stage. A man walks across this empty space whilst someone else is watching him, and this is all that is needed for an act of theatre to be engaged".

Yet our traditions stemming from Ancient Greek Theatre, Western or Chinese Opera with a large audience and elaborate sets and staging, require sophisticated auditoria, vast and complex back staging spaces and technology. This complexity places great demands upon the ingenuity of the architect. The Performing Arts are regarded as one of the great pinnacles of cultural achievement. Grand cultural buildings are usually ranked highly, alongside governmental buildings and places of worship in the urban hierarchy. City planners and politicians have seized upon the special nature of these buildings to give their cities identity.

This book shows is an important record of recent work to demonstrate how architects and their clients have responded to this challenge.

CONTENTS

影院
CINEMA

10

SAUVENIERE CINEMA

22

MULTIKINO STARY BROWAR

28

KRONVERK CINEMA

38

INOX

46

HDIL, KULRAJ-BROADWAY CINEMA AT DREAMS MALL

54

BIG CINEMAS HYDERABAD

60

ODEON WREXHAM

72

CITY CINEMA, OMAN, SOHAR PLAZA

78

PVR CINEMAS SURAT

84

MULTIKINO ARKADY WROCŁAWSKIE

92

GH WHAMPOA CINEMA

98

MULTIKINO PASAŻ GRUNWALDZKI

104

PVR CINEMAS PHOENIX MILLS

112

MBOX CINEMA

剧院
THEATRE

120

OPERA HOUSE OSLO

134

AMPHION THEATRE

146

CORBY CUBE

158

MARLOWE THEATRE

168

SOUTH MIAMI-DADE CULTURAL CENTRE

180

NEW THEATRE IN MONTALTO DI CASTRO

188

BURLINGTON PERFORMING ARTS CENTRE

目录

音乐厅
CONCERT HALL

198
TEATRO MUNICIPAL EN ALMONTE

208
OPERA HOUSE AND POP MUSIC STAGE ENSCHEDE

218
LA LLOTJA THEATRE

228
SHANGHAI ORIENTAL ART CENTRE

236
LE QUAI THEATRE IN ANGERS

246
DEVENTER SCHOUWBURG

256
ONASIS CULTURAL CENTRE

268
AARHUS THEATRE

278
THEATRE AGORA

292
WEXFORD OPERA HOUSE

304
THE PLAYHOUSE THEATRE

316
HARPA-REYKJAVIK CONCERT HALL AND CONFERENCE CENTRE

332
MUSIC HALL EINDHOVEN

342
TAP . THEATRE AUDITORIUM OF POITIERS, FRANCE

354
FREIGHT & SALVAGE

362
INDEX

CINEMA

影院

LOCATION_Liege, Belgium

SAUVENIERE CINEMA

Architect_V+ (Bihain, Decuypere, Hagiwara)
Area_4,286m²
Photographer_ Alain Janssens, V+

Given the complex criteria in the world of cinema and theatre dictated by the incredible technical evolution in this area, as well as the economic considerations, there has been a tendency towards uniformity of design in modern cinema complexes. Faced with these demands the only difference between two modern cinemas, apart from the program, is the welcome and the quality of the public spaces. The project is a piece of lost property (four auditoriums) solid as the soul of a cliff, evoking two enormous bodies squashed into a plot of land too small for them. Opportunistically the working areas are lodged between the different residual spaces. Human traffic bypasses, goes alongside, climbs or crosses, thus offering a succession of stage sets between the object and the town. Nothing on the outside betrays the mystery of the auditoriums. Nomination for Belgian Building Awards, Urbanism prize for the town of Liege, Public Prize of Liège, Winner of Belgian architectural prize, Nomination for Mies Van de Rohe prize, Nomination Archi BAU Europe Awards IN 2009.

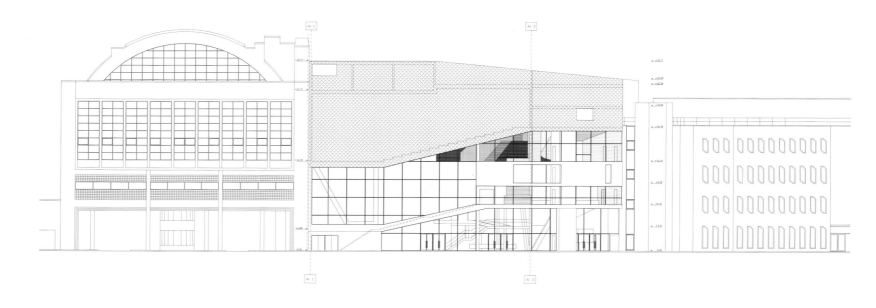

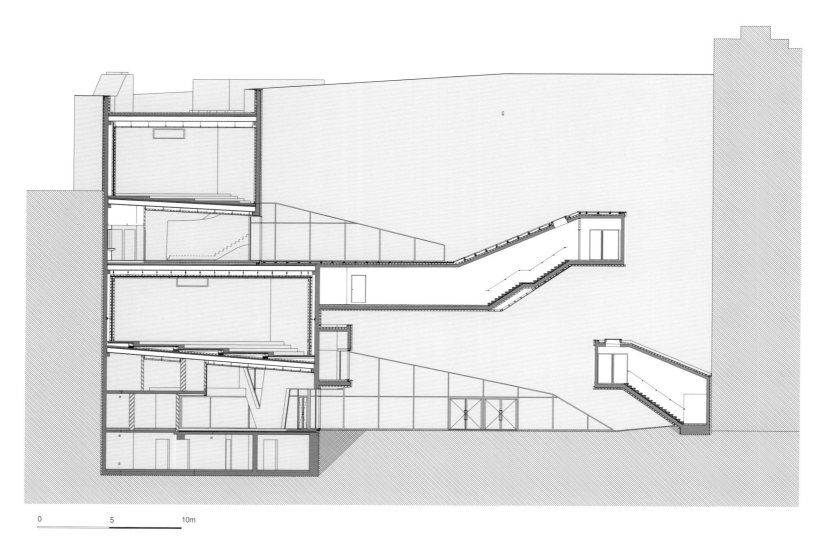

0 5 10m

　　受日新月异的技术革新及经济因素的影响，全球影剧行业已有了复杂的标准，因此，现代复合式影院在设计上已越来越趋向统一化。在这种需求下，除了项目本身之外两个现代影院间的唯一区别便是公共场所的受欢迎程度和质量。该项目是一块废弃的房产（拥有四个礼堂），异常坚实，拔地而起的两座巨大的建筑主体立于相对较小的一块地基之上。工作区域恰巧就设置在剩余的空间内，沿着附近的交通支路，或攀登或交叉，因此，在小镇与建筑物间设立了一系列的阶梯。外观并没有泄露观众席的神秘之处。在2009年此建筑分别获得了比利时建筑奖提名、比利时烈日城都市生活奖、比利时烈日城公共奖、比利时建筑奖、密斯·凡·德·罗奖提名及欧洲建筑BAU奖提名。

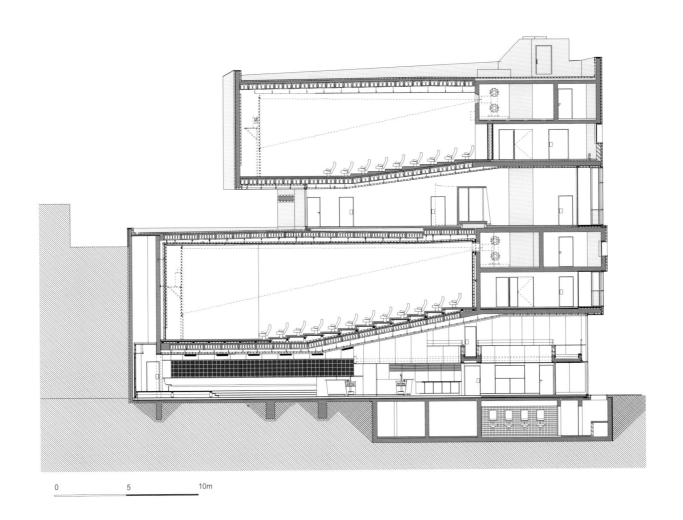

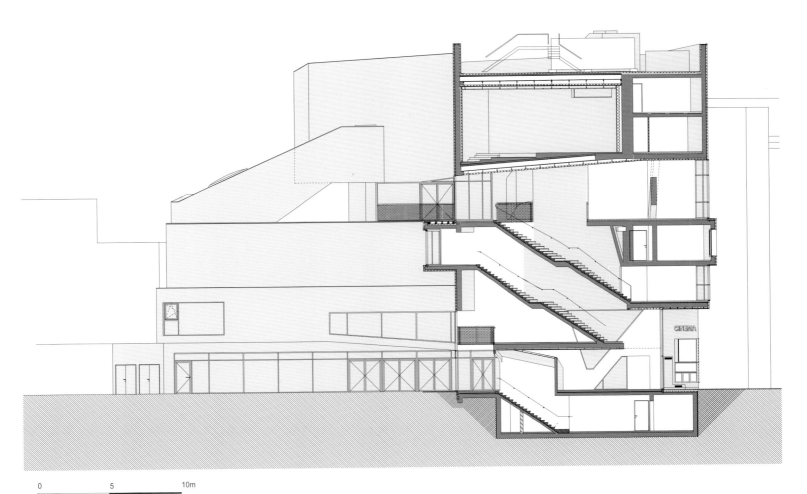

13

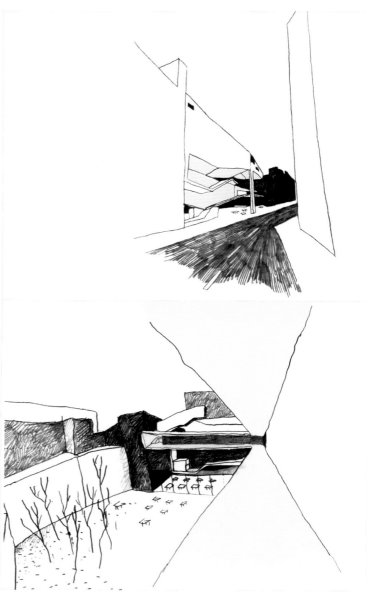

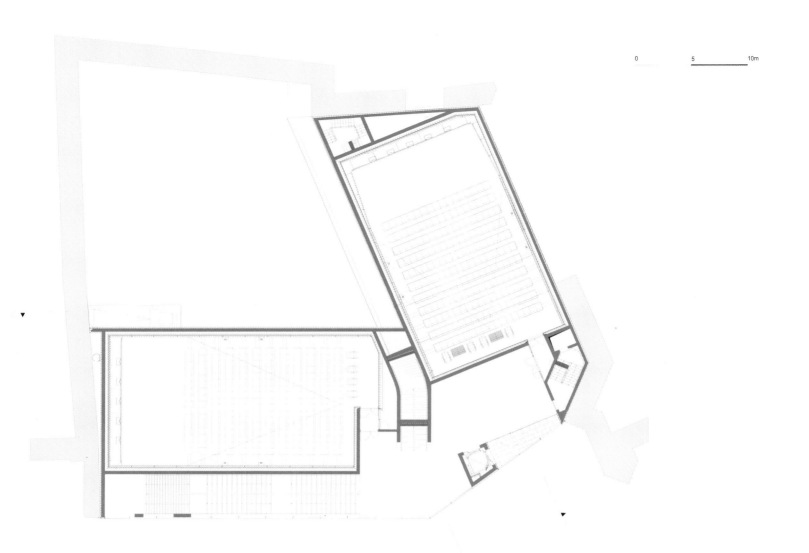

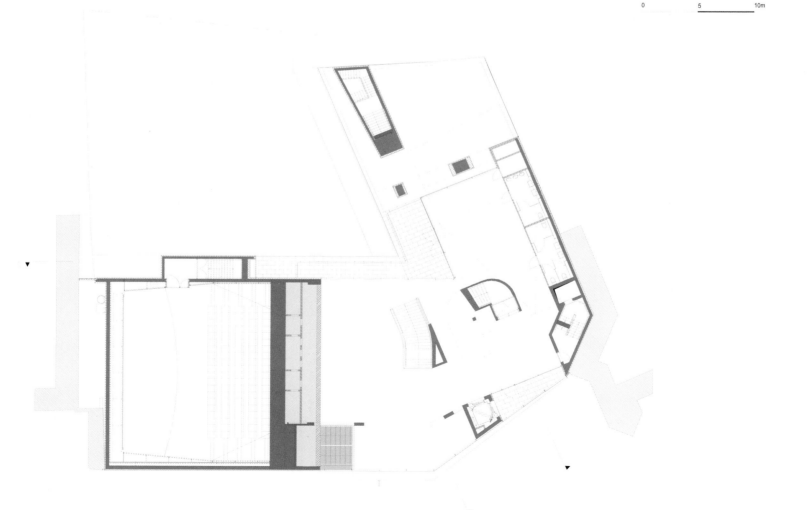

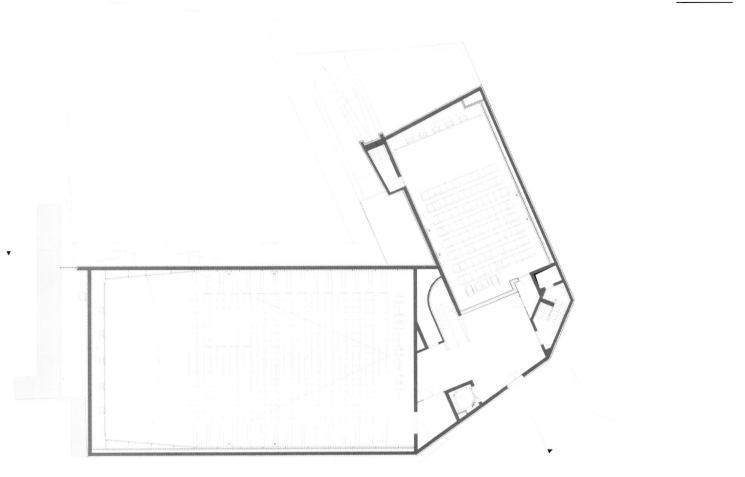
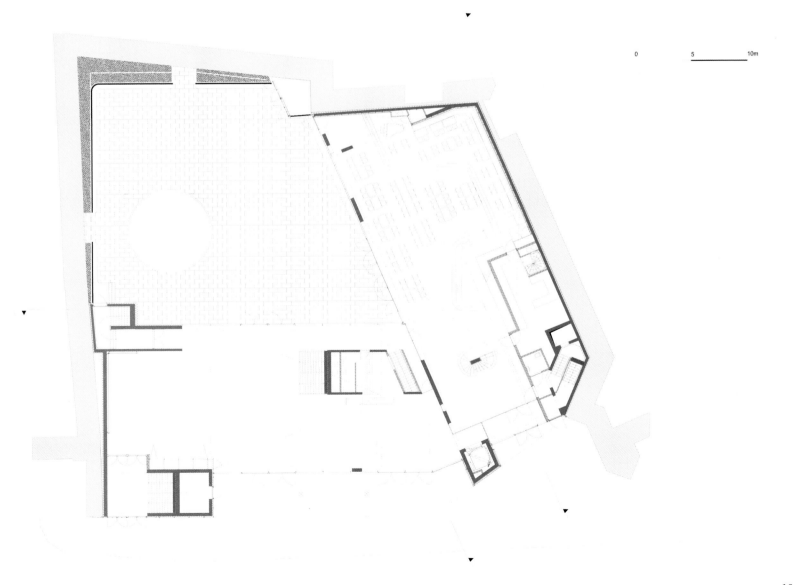

19

LOCATION_+2 Półwiejska St, Poznań, Poland

MULTIKINO STARY BROWAR

Design Company_Robert Majkut Design
Area_640m²
Photographer_Szymon PolaPoznański

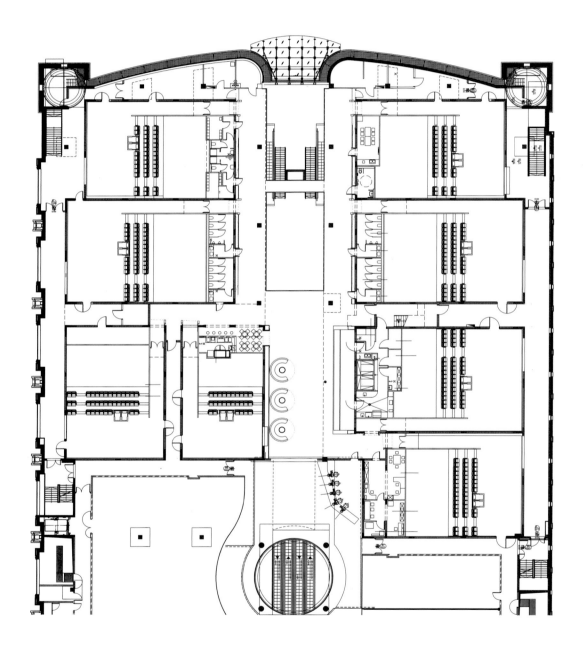

Multikino in Stary Browar in Poznań is yet another example of cooperation between studio and the leader of the entertainment industry in Poland. Thoughts about advantages, which the concept of Stary Browar can bring, and what the Multikino can convey, were fundamental in the project. Stary Browar is a special place, combining commerce and art, the history and the present, uniqueness and everyday life. Set of arrangement means used in the cinema emphasize the individualism of the whole concept of Stary Browar. The connection was founded in the stylistic overtone: post-industrial retro style of the shopping centre was reflected in retro convention of Multikino cinema. The interior surprises with its décor and technology. Multikino in Stary Browar is the first place in Poland where digital projectors, ensuring the best sound and image quality, were installed. The arrangement of foyer rooms and cinema corridors were created in retro science-fiction style. Elements of arrangements' forms and materials used here correspond to romantic science fiction cinema in its earliest days – e.g. "Star Wars" directed by George Lucas. Corridors leading to the cinema halls imitate structure of the spacecraft interior – characteristic lightning, hatches, portholes, rough texture of walls and ceiling. All of this intensifies the magic atmosphere of the cinema and Stary Browar.

位于波兹南 Stary Browar 的 Multikino 影院是工作室与波兰娱乐业领头羊的另一个合作项目。Stary Browar 的理念应该是什么，Multikino 影院能够传达些什么，诸如此类的想法是本次设计项目的根本。Stary Browar 是个特别的地方，它汇集了商贸和艺术，融合了历史与现代，体现了独到与平凡。影院中使用的布局方式强调了整个 Stary Browar 个性化的理念。购物中心与影院间的联系在设计风格中得到体现：购物中心的后工业时代的复古风格在 Multikino 影院的复古结构中依稀可见。室内设计的装潢与技术令人惊叹。位于 Stary Browar 的 Multikino 影院是波兰首家装有数码播放设备的影院，可确保影片拥有最优质的音效和画面。休息室的布局和影院走廊的设计都采用科幻风格。布局形式和使用的材料都符合早期浪漫主义科幻电影（比如乔治·卢卡斯导演的《星球大战》）的要求。通向影院大厅的走廊的设计仿照了宇宙飞船的内部格局——别致的灯光、舱口、舷窗、粗糙质地的墙壁和天花。所有这一切都强化了影院和 Stary Browar 的魔幻氛围。

LOCATION_Moscow, Russia

KRONVERK CINEMA

Designer_Robert Majkut
Design Company_Robert Majkut Design
Area_1,045m²
Photographer_Andrey Cordelianu

Kronverk Cinema project was realized for one of the leading cinema network on the Russian market in one of the cinemas in Moscow. The new standard created by Robert Majkut Design allows to be repeated both in new and existing cinemas, as a coherent and consistent solution of the spatial organization and visual identification.

The concept is realized as a smooth development of the logotype's shapes and its color balance. All spatial geometric forms present in this project are the consequence of a certain order of lines, which comes from the trademark – most of all from the symbol of crown. Despite the consistent use of color and geometry in the interiors of Kronverk Cinema, each zone is characterized with a distinct and clearly codified aesthetics. The entrance to the cinema, modern and geometric, leads to the lobby area – a very characteristic tunnel, divided into spatial zones with seat modules. The ceiling of the lobby is a set for the magnificent play of light formed in a subtle pattern. Alcohol Bar is to give the impression of saturated, patterned spaciousness. Completely monochromatic, yellow VIP Bar provokes a feeling of immersion into a unique and energizing space, decorated with figurative motifs. In opposition, the VIP Lounge, with its soothing tones of black and purple, creates a climate of intimacy, the feeling of being incognito in an elegant room decorated with precious materials. An unusual solution is the white corridors, strongly contrasting with the dark cinema halls. Thus, each room in this facility is clearly defined, both functionally. This design in a modern way refers to the Russian decorativeness. The patterns and forms applied in the project are contemporary interpretation of motifs known from other eras or traditional ornaments. The best example is the ceiling in the lobby with LED lights arranged in a delicate, lacy pattern. The power of expression of this interior defines the modern eastern splendor, based on the latest technological achievements.

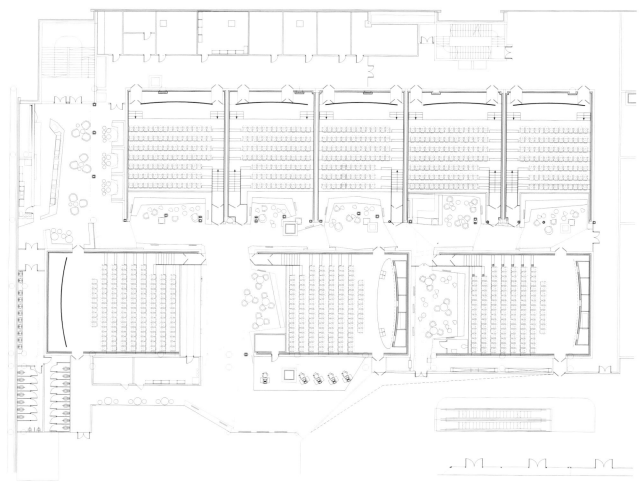

• Floor Plan

　　Kronverk 影院是莫斯科众多影院之一，拥有俄罗斯领先的影院系统。Robert Majkut 设计公司的新设计标准是使空间构架和视觉效果连贯持续并适用于新老影院。

　　设计概念的实现得益于标识外形及其颜色的均衡发展。设计师从影院商标——一个皇冠中获取灵感，该项目展示的所有空间几何造型都是影院商标里线条排列的产物。尽管 Kronverk 影院在室内使用的是相同的颜色和几何图形，每一个区域却又各具特色，拥有迥然不同的风格。富有现代感的几何图形装饰的影院入口通向休息区——一个独具特色的隧道，被座位分割成多个小区域。休息区的天花板被灯光装饰成精美的图案，营造出了不可思议的效果。酒吧给人饱满开阔的印象。黄色的贵宾酒吧色调单一，让人仿若进入了一个用生动图形打造的非凡、动感的空间。相反地，贵宾室的设计采用了凝重的黑色和紫色，用上等材料装潢出优雅的环境，为顾客营造出一种私密氛围，一种"隐士"之感。白色走廊与影院的黑色墙壁形成强烈的反差，因此，影院的每个空间都色彩鲜明、功能独特。这种走现代路线的设计和俄罗斯的装潢风格相契合。该项目应用的图形和造型都是对来自于其他时期或其他传统装饰图案的现代诠释，最好的例子就是休息室的天花，LED 灯被巧妙地排列起来，呈现出花朵的图案。此种室内设计，通过运用最新的科技成果，淋漓尽致地表现出了现代东方风格的华丽大气。

LOCATION_Thane, India

INOX

Design Company_Arris Architects Pvt Ltd
Area_1,300m²
Photographer_INOX Leisure Ltd, Ajinkya Manohar

The principle attempt for the design was to break away from the conventional format of following the slopes of the cinemas, in turn bringing about a new brand identity for INOX. Cinema panel design interprets the side façade as a wall piece that would give a pleasant visual experience for the patrons at time of entering the cinema and intermission, while at the same time not bold enough to distract during movie watching. The wall skin is a blend of different colors in a drape form, at the same time making the design more time and cost effective. Design revolves around the concept of contemporary, vibrancy, playfulness, and optical illusions. The layering of colors juxtaposed within the main colors lends the skin a varied appearance, which is vivid and three dimension from close-up and appears homogeneous and flat from far. The colorful stripes design pattern is aptly concise in the ribbony drapes which are held within the brown fabric paneling. Interior space of concession lobby is an extension of the same concept though more like a prologue to it. The lobby space is inscribed in an inclined elliptical shell, increasing its effective volume & detaching itself from the direction expression of structural slopes. The use of mirrors on beams makes the huge structural beams disappear & brings lightness to the space. It uses the identical stripe pattern in a more subtle way. Wall cladding in vitrified tiles with a monochromatic pattern identifies itself with cinema design. Concession counter shell displays a white shell running throughout the linear interior space and caving in at places with bright yellow cross sections for counters and auditorium entries. Flooring of silver travertine along with inlays of white agglomerated marble flows through the linear space. The same lines extrude themselves to form seating enclave. Front graphic wall & pattern in the ceiling add vibrancy to the space. The entire space displays various elements each expressing their individuality, however bound together by a central concept and detailing, thus forming an integral space with variety of experiences justifying its concept.

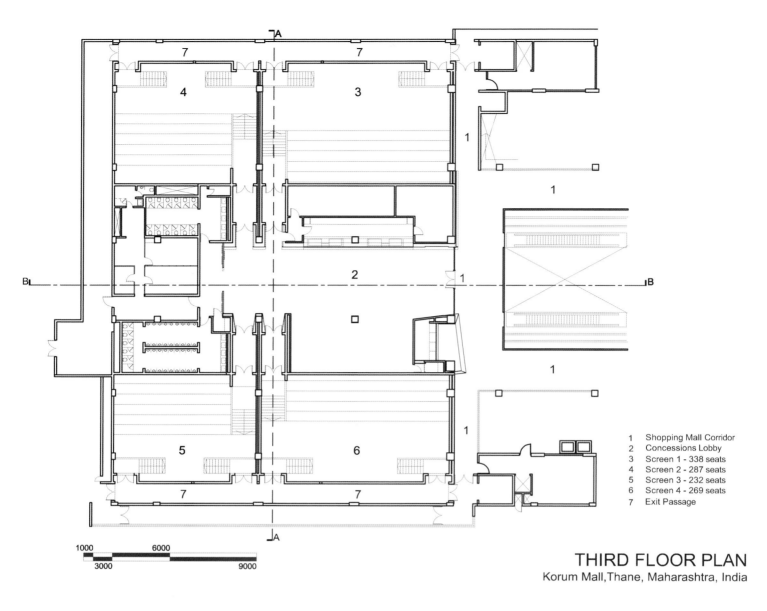

1 Shopping Mall Corridor
2 Concessions Lobby
3 Screen 1 - 338 seats
4 Screen 2 - 287 seats
5 Screen 3 - 232 seats
6 Screen 4 - 269 seats
7 Exit Passage

THIRD FLOOR PLAN
Korum Mall, Thane, Maharashtra, India

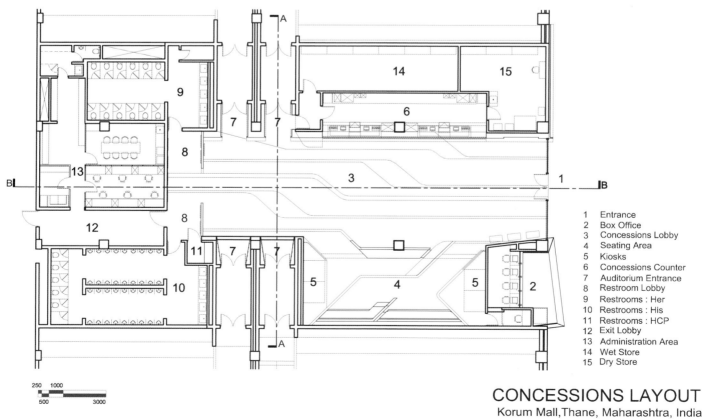

1 Entrance
2 Box Office
3 Concessions Lobby
4 Seating Area
5 Kiosks
6 Concessions Counter
7 Auditorium Entrance
8 Restroom Lobby
9 Restrooms : Her
10 Restrooms : His
11 Restrooms : HCP
12 Exit Lobby
13 Administration Area
14 Wet Store
15 Dry Store

CONCESSIONS LAYOUT
Korum Mall, Thane, Maharashtra, India

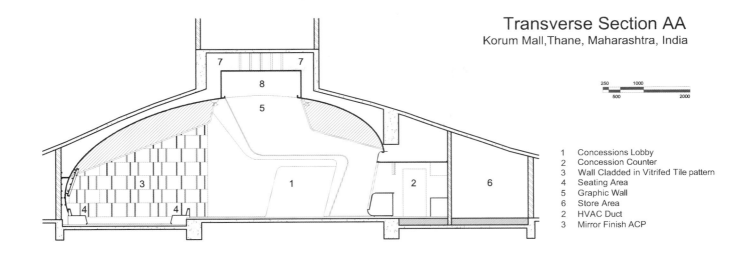

Transverse Section AA
Korum Mall, Thane, Maharashtra, India

1 Concessions Lobby
2 Concession Counter
3 Wall Cladded in Vitrifed Tile pattern
4 Seating Area
5 Graphic Wall
6 Store Area
2 HVAC Duct
3 Mirror Finish ACP

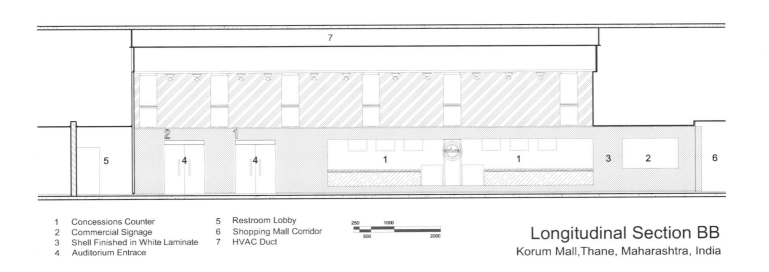

1 Concessions Counter
2 Commercial Signage
3 Shell Finished in White Laminate
4 Auditorium Entrace
5 Restroom Lobby
6 Shopping Mall Corridor
7 HVAC Duct

Longitudinal Section BB
Korum Mall, Thane, Maharashtra, India

这项设计的主要目的是脱离传统影院的斜坡式造型，赋予INOX新的品牌形象。经设计，影院侧面的墙片能让观众在进入影院和幕间休息时拥有愉快的视觉效果，同时侧墙也不会因为过于醒目而分散观众观看影片时的注意力。彩色的墙皮呈褶皱状，使设计更加经济高效。设计围绕现代、活力、娱乐和幻觉展开，颜色层次与主色彩相融，增添了墙皮外观的多样性。近距离看，生动而有三维效果；远观时则是二维的平面效果。彩色的条纹图案巧妙地突出了隐藏在棕色墙体中的彩虹图形。贵宾室的内部空间与其说是本次设计主题的序曲，不如说是对主题的延伸。休息室形状似一个椭圆形的贝壳，这样的设计增强了室内的音效，同时也使这个空间摆脱了斜坡式结构的表现形式。横梁上安装的镜子使巨大的横梁在得以遁形的同时又将光线引入影院。设计巧妙地运用了同样的条纹图案达到了奇妙的效果。琉璃瓦在墙壁上拼出一个单一的图案与影院设计完美契合。线性的室内空间里，明黄色交叉区标示出前台和礼堂入口，其间坐落着形似白色贝壳的贵宾室。银灰洞石地板与镶嵌于其上的白色人造石穿流在这个线性空间中。线条凸起的部分则成为观众席。前方富有动感的墙体和天花板图形为影院增添了生机。整个空间展示出了多种元素，每种元素都有其独特性，却又通过一个核心主题和设计细节紧密地联系在一起，即以不同的体验搭建起一个整体空间。

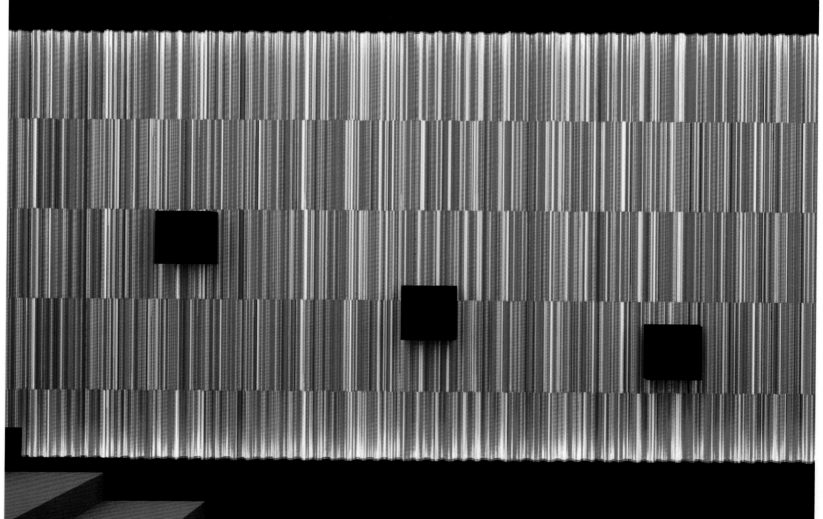

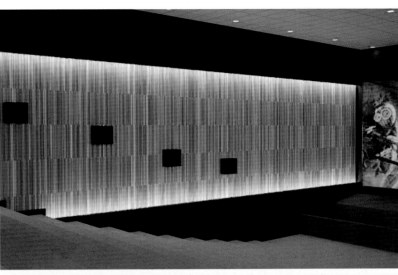
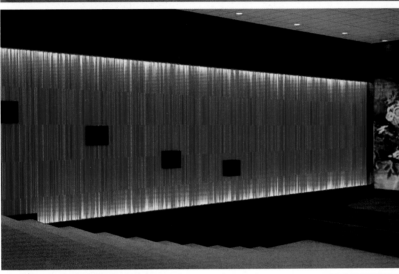
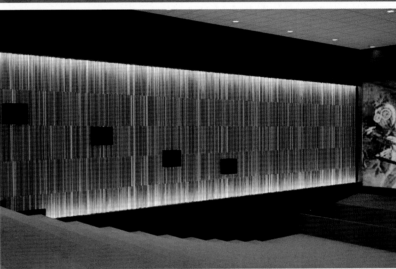

LOCATION_Mumbai, India

HDIL, KULRAJ-BROADWAY CINEMA AT DREAMS MALL

Designer_Mr. Eranna Yekbote
Design Company_ERA Architects
Area_6,159m^2
Photographer_Mr. Girish Patil

HDIL, a major real estate company launched its venture Dreams Mall in a prime location and busy suburb of Mumbai, with Broadway as its cinema brand. It is a six screen cinema with a spacious lobby and VIP lounge facility. The concept for cinema lobby was inspired from facets of a diamond. The dynamic movement and play of light on different facets of diamond has been interpreted in the ceiling of the cinema lobby. The color is a light blue contrasting with white with seating niches carved in dark blue. The concept is further taken into the auditoriums, on their side walls and ceiling, with varying color schemes in each of them. The Gold Class Lobby is on a totally different theme with black as dominant color giving it a high class look. Ceiling is designed with plaster of Paris cubes of uneven heights suspended from ceiling creating an interesting ambience. Overall this project is experimentation with architectural form and interplay of light and color.

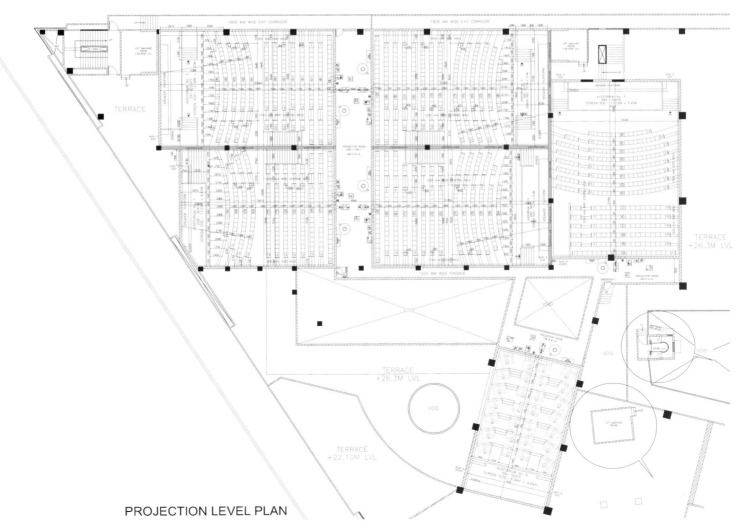

PROJECTION LEVEL PLAN

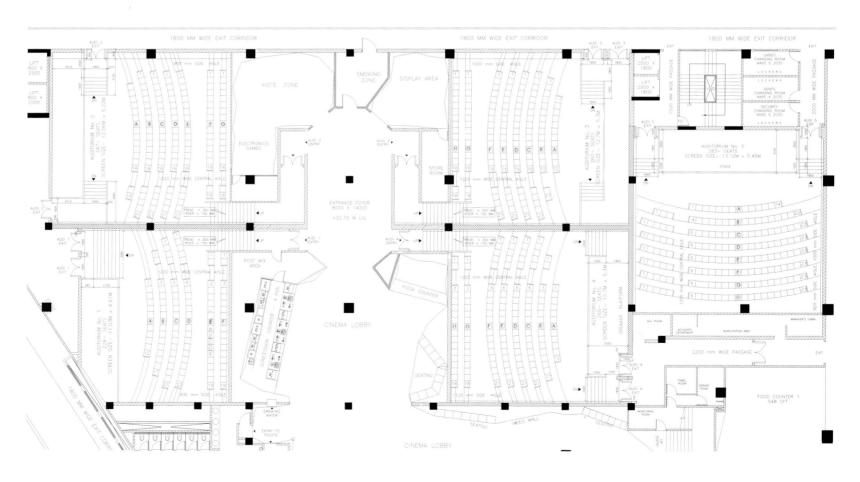

在孟买郊区繁忙的黄金地段，大型房地产公司 HDIL 投资建造了梦幻商场，以百老汇作为其公司的电影品牌。影院有六个屏幕、一个宽敞的大厅和 VIP 休闲设施。大厅的设计灵感来自于钻石的切面。影院大厅的天花设计展现了光束照射在钻石各个切面所形成的动感效果。浅蓝色与白色带有深蓝色雕刻图案的壁龛形成鲜明的色彩对比。设计进一步地考虑到了听众席的效果，侧面墙和天花板实行了不同的配色设计方案。Gold Class 大厅则展示了完全不同的主题，背景以黑色为主突显出其高端的品质。悬挂在天花板上的参差不齐的熟石膏立方体使天花板的设计显得尤为有趣。总体上，此项目是检验建筑形式、光与色彩间相互作用的一项实验。

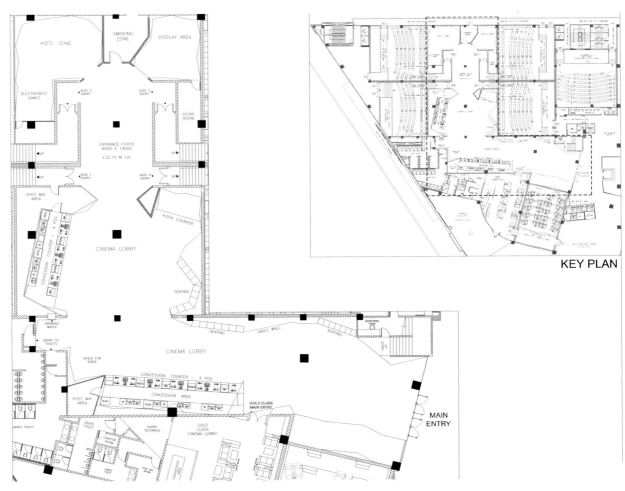

KEY PLAN

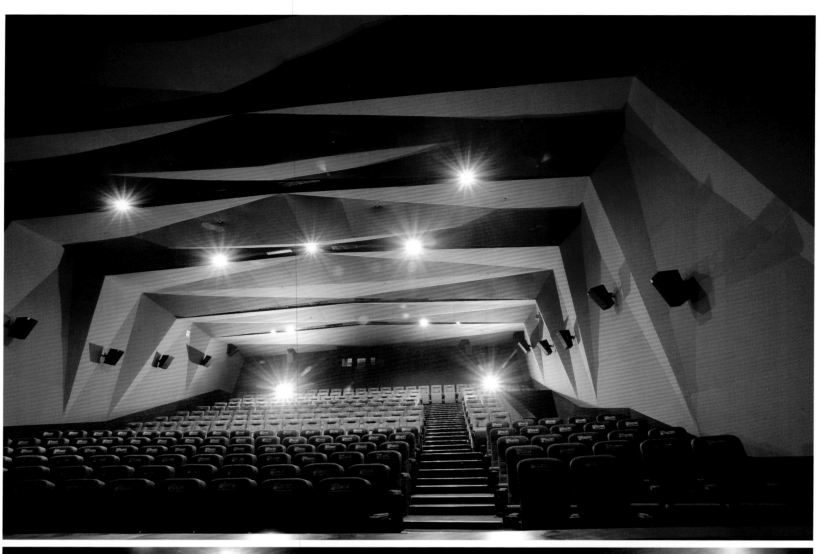

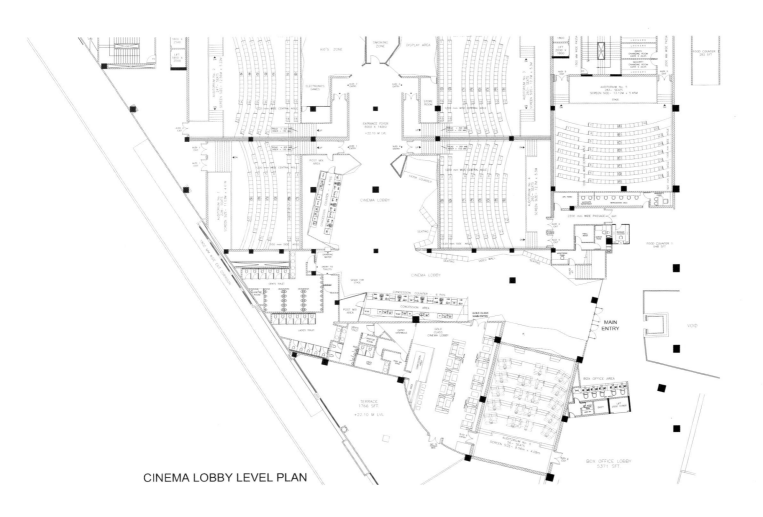

CINEMA LOBBY LEVEL PLAN

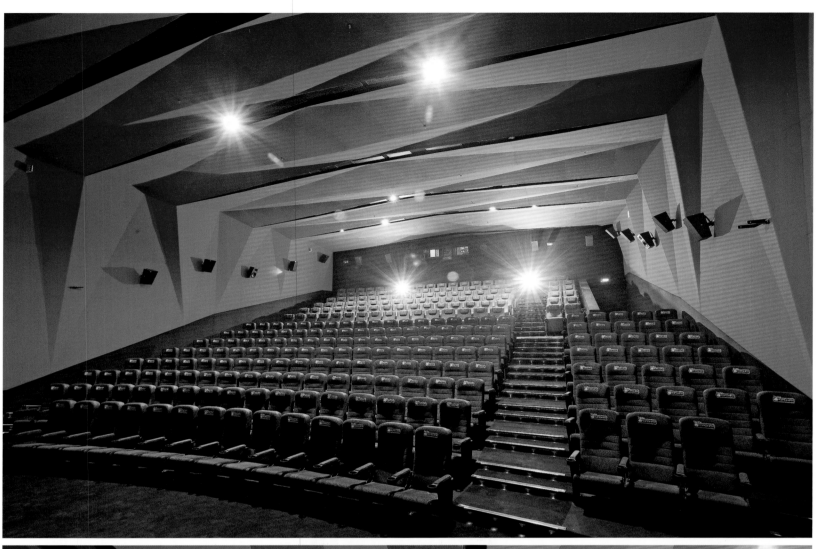

LOCATION_Hyderabad, India

BIG CINEMAS HYDERABAD

Chief Designer_ Mr. Eranna Yekbote
Project Designer_ Mr. Vikas Jain
Design Company_ ERA Architects
Area_ 3,378m²
Photographer_ Mr. Girish Patil

Big Cinemas Hyderabad was completed in the year 2007 and is located in one of the busiest cities of India. The cinema spreads around an area of 3,378m² and has four screens with 1,400 seats in total. The concept of the lobby evolved around ribbon which envelops whole of the space creating an engaging environment. The pattern begins symbolizing a red carpet at the entrance leading to concession counter, going up in the ceiling, turning down on the walls terminating as seating and leading to auditorium entry doors. Red is the predominant color contrasted with white and brown. The Auditoriums have a red and dark blue color palate with subdued lighting and efficient acoustics. The simplistic lines and lighting create an overall soothing ambience with flooring in subdued beige bringing out the design elements. This project truly evolves from initial concept development into a high-class design.

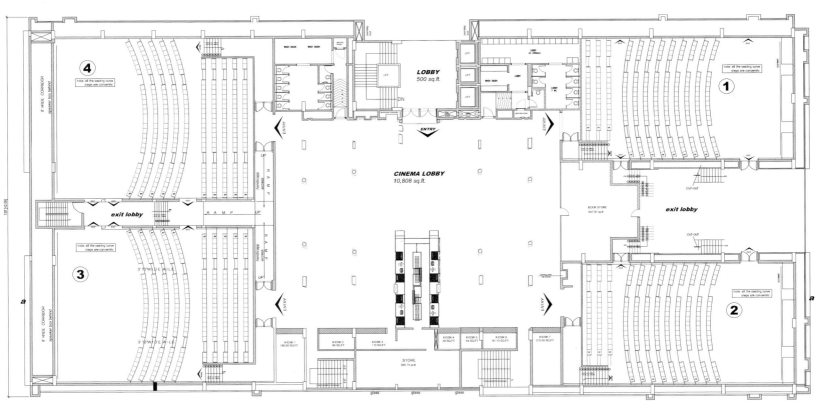

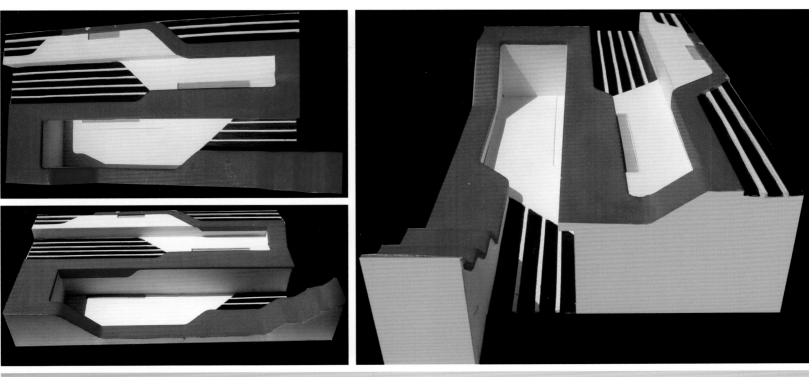

海德拉巴大影院位于印度最繁忙城市之一的海德拉巴，于2007年竣工。影院占地3378平方米，共有四个大屏幕和1400个席位。大厅的设计理念源于彩带，笼罩着整个空间营造出迷人的氛围。图案起于入口处象征着红毯一直通向接待柜台，再向上延伸到天花板，转向墙壁，最终化为席位通向观众厅的入口。与白色和棕色相对比，红色是主色调。观众厅红色和深蓝色的色彩搭配与柔和的灯光及优质的音效融合在一起。极简的线条和光束与柔和的米黄色地板凸显出的设计元素一同营造出舒缓的氛围。这个项目从最初的概念真正演变成为了一件高水平的设计作品。

LOCATION_Wrexham, North Wales, UK

ODEON WREXHAM

Designer_Glyn Mellor
Design Company_NBDA Architects
Area_3,200m²
Photographer_G Mellor

In 2007 NBDA Architects were appointed by Odeon Cinemas to design the new 8 screen multiplex cinema at the new Eagles Meadow Shopping Centre development in Wrexham.

The cinema entrance is from an external walkway at the first floor level of the retail complex. The minimal entrance frontage was designed to fit with the proportions of the adjacent retail units. The cinema entrance also incorporates a Costa Coffee Bar and vertical circulation to the cinema level above. The cinema screens are located either side of the concourse. The irregular shape of plan allows screens of varying sizes and capacities to be created. The screens range in size from 105 seats up to 220 seats to produce a total capacity of 1,450 seats including wheel chair positions. The screens are all designed to fill the front of the auditoria with moveable side masking in black fabric. The masking allows the visible screen proportion to be adapted for different film formats. Toilets and ancillary rooms are also located either side of the concourse using the space below cinema seating to make best use of the floor space.

2007年，Odeon影院委托NBDA设计师事务所在雷克瑟姆的"鹰坪"购物中心设计一个新的八屏多元化影院。

影院入口是从零售大楼一楼的外部人行道延伸进来的。最小入口的尺寸与其毗邻的零售店铺比例协调。影院入口设有Costa咖啡店，可以由此直接进入楼上的影院。沿着广场的一侧安装着整齐的影院屏幕。建筑场地呈现不规则形状，因此，设计师设计出了尺寸大小各不相同的屏幕。屏幕尺寸小到可供105人观赏大到可供220人观赏，整个空间包括轮椅座位在内共可容纳1450人同时观赏。经过设计，所有的屏幕都安装在观众席前方，可移动的一面隐藏在黑色的布景中。由于屏幕一侧可以隐藏，影院的屏幕便能适用于不同的电影模式。洗手间和辅助间被设计在在广场的另一边，充分利用了影院座椅下方的楼层空间。

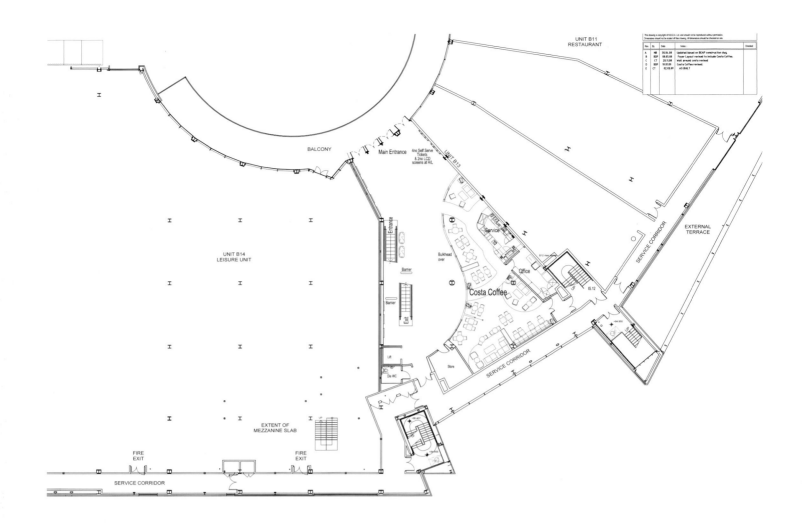
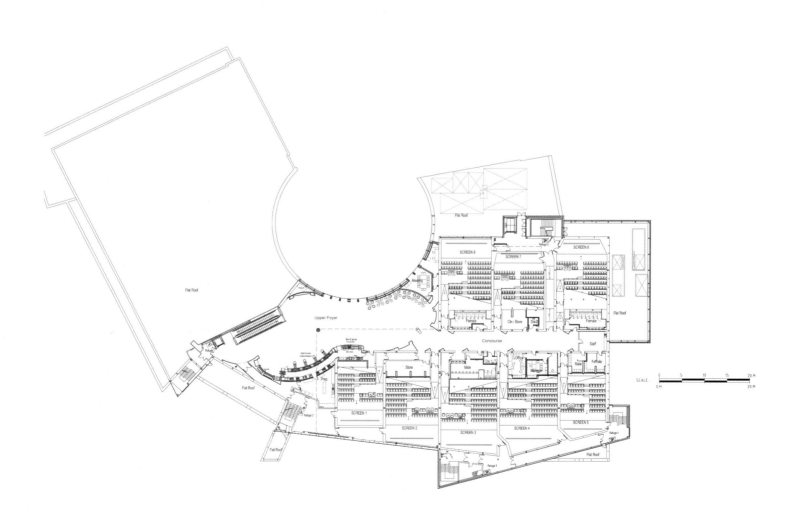

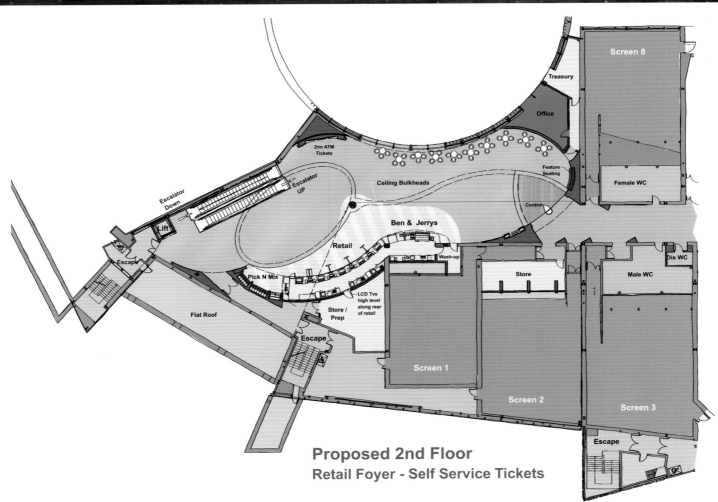

Proposed 2nd Floor
Retail Foyer - Self Service Tickets

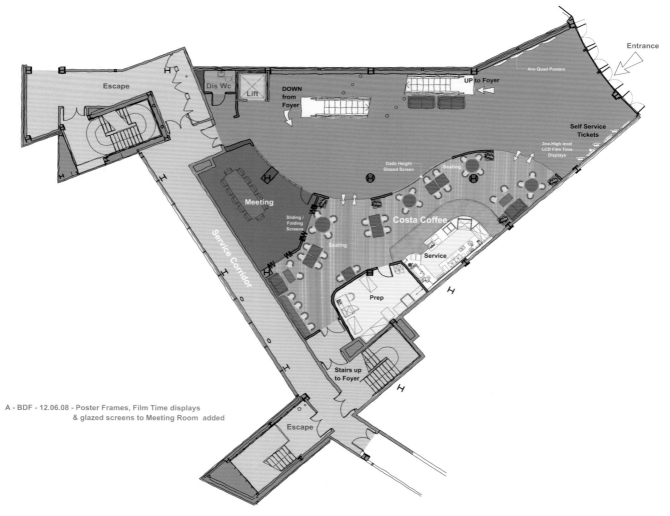

A - BDF - 12.06.08 - Poster Frames, Film Time displays & glazed screens to Meeting Room added

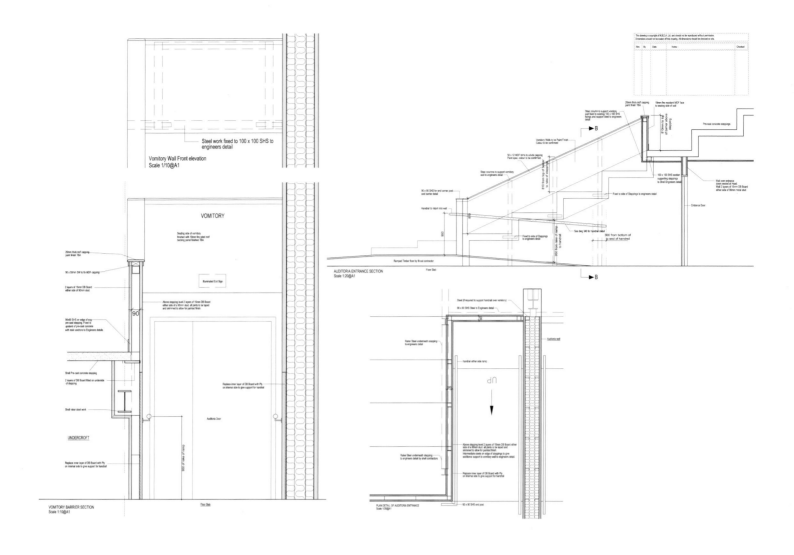

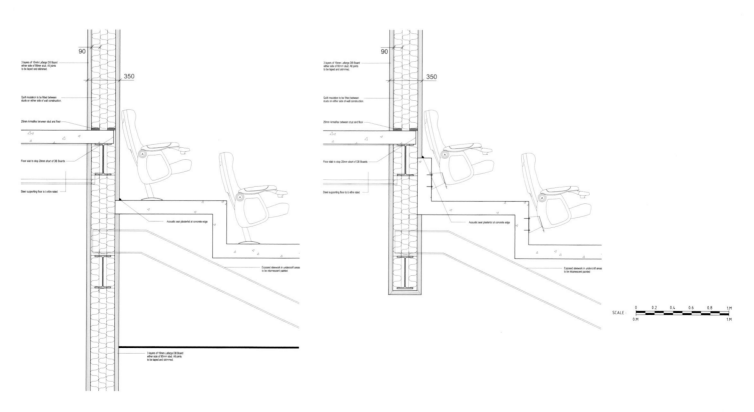

SECTION THROUGH REAR WALL OF AUDITORIA
WITH REAR WALL TAKEN DOWN TO GROUND

SECTION THROUGH REAR WALL OF AUDITORIA WITH
UNDERCROFT AREA AS PART OF FOYER

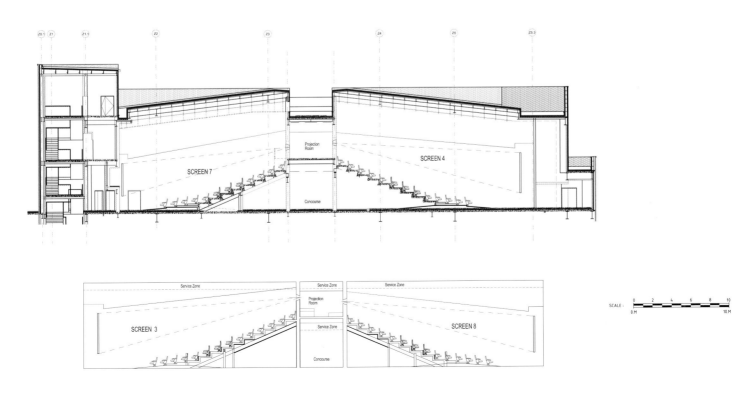

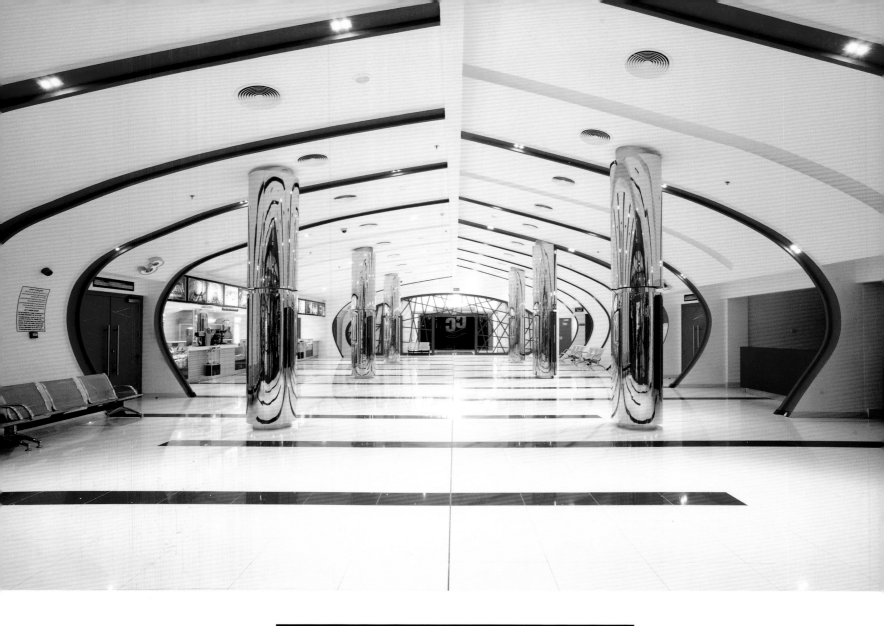

LOCATION_Sohar, Oman

CITY CINEMA, OMAN, SOHAR PLAZA

Chief Designer_Mr. Eranna Yekbote
Design Company_ERA Architects
Area_7,497m²

This is a standalone cinema project in the town of Sur, Oman. It seats around 800 people and has screens placed back to back with a single projection room. The concept of cinema lobby is simplistic, yet with a grand feeling due to use of arcs which give a fluid quality to space. Architecture of Oman is noted by its arches and delicate "jaali" work. Both these elements have been interpreted and abstracted in the overall cinema lobby design. The entrance to lobby is an abstracted "jaali" bright orange in color, marking a vibrant entrance. Concession counter is in subdued yellow hues and red color. Columns in the lobby are clad with stainless steel and mirror the array of arches creating a surreal effect. Overall color palate of the lobby is a combination of white and orange paint finish giving it a crisp look. Auditorium design reflects the abstraction of arch on the side wall panels of the auditorium carrying ahead the concept in the lobby. The outer elevation of the building is designed with local materials and traditional elements. Attention to symmetry, repetition of elements and proportions are some of the marked attributes of ancient architecture and this lobby design reflects the same and is a perfect example of taking inspiration from the old and interpreting it in a modern way.

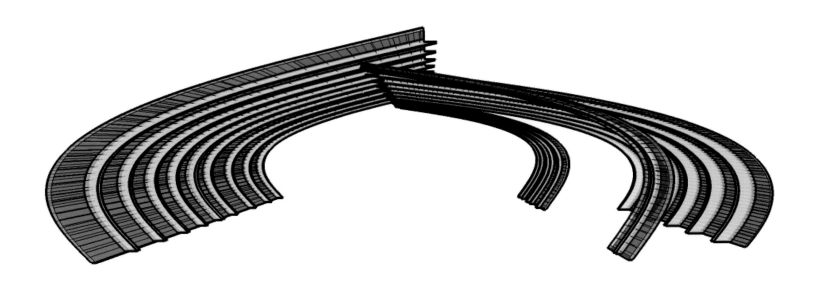

这是一个在阿曼苏尔城的独立影院项目。影院可容纳800人，屏幕背靠背地放置在独立放映室中。大厅的设计理念是追求极简风格，弧光灯的使用营造出空间流动感给人以宏伟的感觉。阿曼的建筑风格以拱形和精美的装饰图案而著称。这两种元素均被诠释和抽象化地运用在整个电影院大厅的设计中。大厅入口处以明亮的橘色抽象化精美图案来装饰，营造出充满活力的氛围。接待柜台以黄色和红色色调为主。大厅矗立着不锈钢的圆柱，映射着拱形阵列营造出超现实主义的效果。大厅的整体色调为白色和橘色的搭配，给人清新之感。观众席的设计反映出观众席侧面墙板拱形设计的抽象性，继续推进了大厅设计的理念。建筑外围立面的设计融入了当地材料和传统文化的元素。对对称性、元素和比例的重复使用是古代建筑的显著特征，大厅的设计不仅反映出了这些特征，还是从古代建筑中找寻创作灵感并以现代的方式给予诠释的最好的示例。

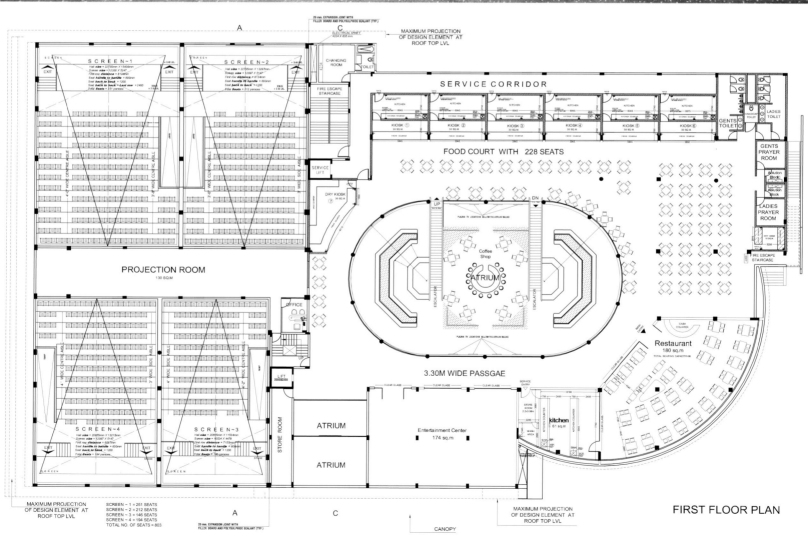

FIRST FLOOR PLAN

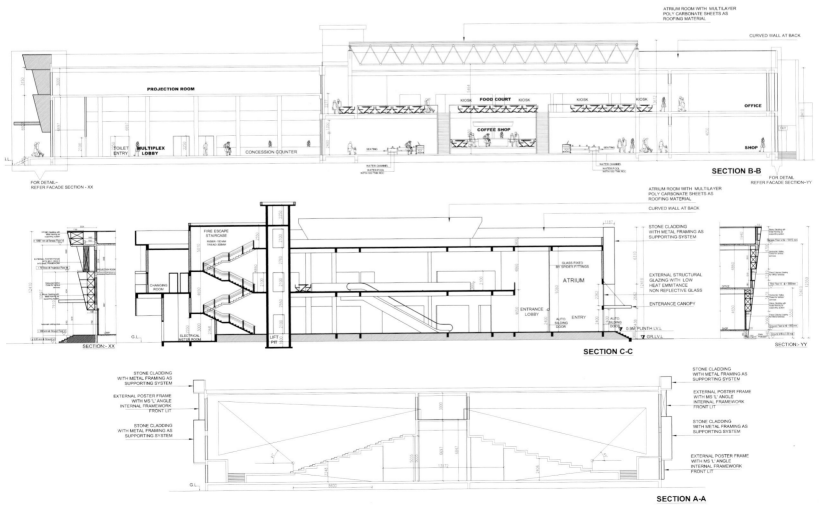

LOCATION_Surat, India

PVR CINEMAS SURAT

Design Company_ ERA Architects, Mumbai and ROR Studio
Area_ 6,503m²
Photographer_ Mr. Girish Patil

PVR Surat was completed in the year 2009 and is an eight screen cinema project housing about 1,780 seats. Surat, known as "DIAMOND CITY", is famous for jewelry, rich culture and tradition. The cinema is about 6,503m² in area and has a spacious lobby space. The concept, for cinema lobby design takes inspiration from diamonds and its glimmering facets. The ceiling is designed like cuts on a diamond face and is studded with LED lights to give that shimmering effect of a diamond. The ceiling gradually slopes towards concession counter which is highlighted with alternating red and yellow back-lit color bands, which also is a focal point from the main entry to cinema lobby. Portion of the ceiling where the height is more, is used for hanging panels of different lengths with colored films having photos of famous actors. A medley of purple, green and red shades is used and light washes on them from the top creating an interesting spectacle. Red is used extensively on columns and side wall of the lobby and flooring is of subdued beige and brown. Poster boxes, LCD's are fixed on side walls as well as on columns projecting the latest updates of upcoming movies. Lobby seating is an interesting pattern of staggered cubes draped with colored fabrics. Overall the lobby takes cue from the concept of a diamond and translates it in terms of light, color and form creating an interesting ambience.

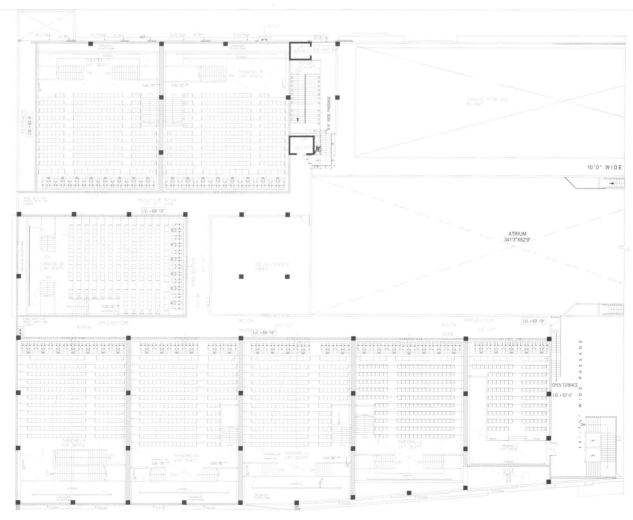

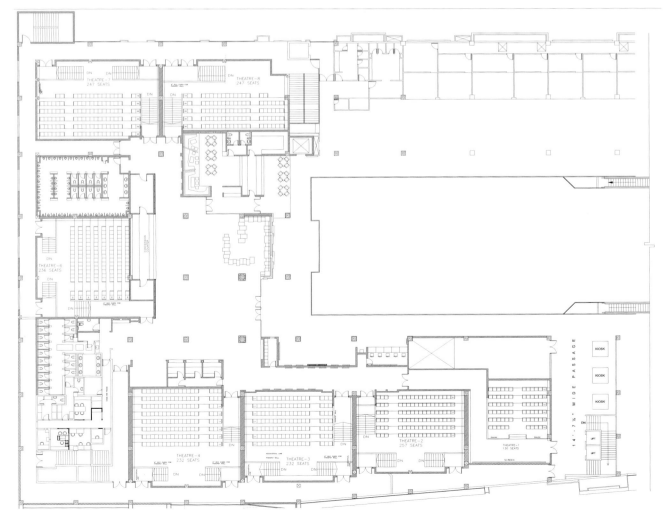

81

苏拉特 PVR 影院竣工于 2009 年，院内有八个屏幕，1780 个观众席位。苏拉特享有"钻石之城"的美誉，因其精湛的珠宝加工技术、深厚的文化底蕴及悠久的传统而闻名。影院占地面积约 6503 平方米，内设宽敞的大厅。影院大厅设计理念的灵感源于钻石及其闪亮的切面。天花板设计得宛如钻石切面，内置的 LED 灯管为这些"切面"营造出钻石般闪闪发光的效果。在红黄交替的背景灯光的凸显下，天花板逐渐向接待柜台倾斜，这一段区域同时也是影院由入口至大厅的亮点所在。天花板相对高出的一部分被用于悬挂不同长度的嵌板，嵌板上印有知名影星们的彩色照片。影院以紫色、绿色和红色为混搭，加之灯光从顶端洒下，营造出一种有趣的氛围。红色用于对圆柱、大厅边墙进行视觉上的延展，地板则是淡淡的米黄色与棕色。液晶显示的海报栏安装在边墙里，最新上映的电影预告片也投映在圆柱上。大厅座椅的形状很有趣，是套着彩色织布的交错放置的立方体。整个大厅以钻石为主题，设计师将这一主题通过灯光、色彩和造型生动地展现出来，营造出一个趣味空间。

LOCATION_2–4 Powstańców Śląskich St, Wrocław, Poland

MULTIKINO ARKADY WROCŁAWSKIE

Design Company_ Robert Majkut Design
Area_ 1,970m²
Photographer_ Szymon Polański

Another stage of cooperation with Multikino brought the creation of interior design for the new cinema object. The whole interior is characterized by natural smoothness and softness of lines, very typical emblem, recognizable feature of style. They refer to the human body movement and its smoothness. The effect of unity of the space is emphasized with shiny surfaces of walls and ceilings, columns' lines and linear lightning. All forms harmonize with each other and permeate gently. The interior is defined with four strong colors: white, red, blue and pink, which were used to mark out the box-office and bar zones. The whole design is a combination of various modern, smooth shapes of chairs, lounges and bars, that create an effect of spaciousness and flow of forms.

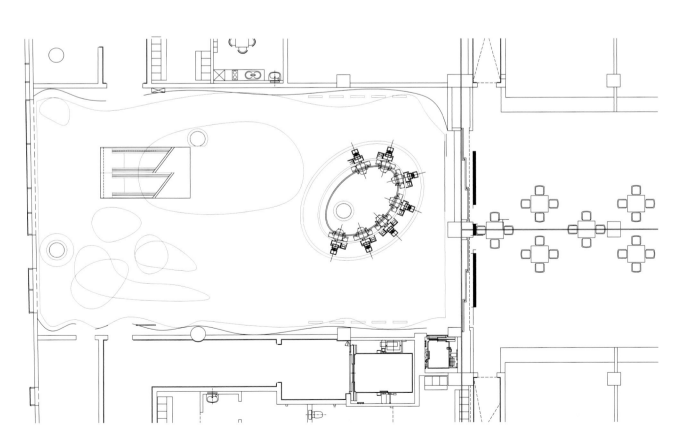

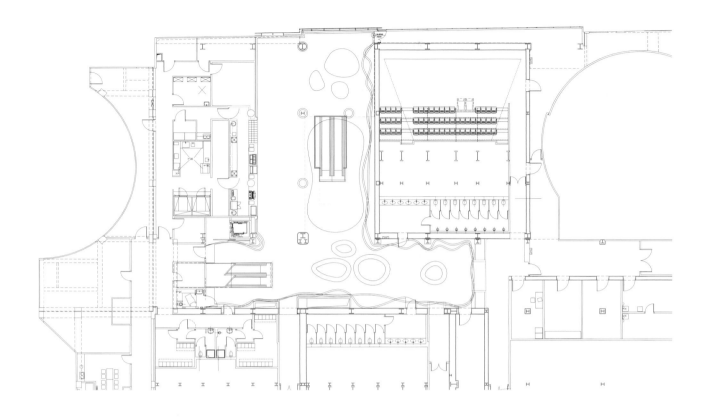

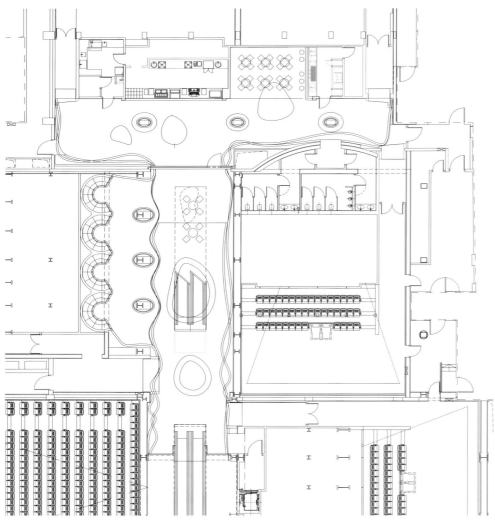

和 Multikino 合作的另一个阶段是新型影院项目的室内设计。整体室内线条自然柔和，非常具有典型的象征意义，风格独特。设计师们参考了人体活动行为及其柔韧性进行设计。整体空间效果体现在闪烁的墙体和天花板、直线型立柱及线性灯光上。所有这一切相互间完美契合，互为衬托。室内有四种主打色：白色、红色、蓝色和粉色，通过不同的颜色突显出售票处和酒吧。整体设计将多种充满现代感、柔和造型的座椅、休息区和酒吧结合在一起，营造出一种开阔、流动的效果。

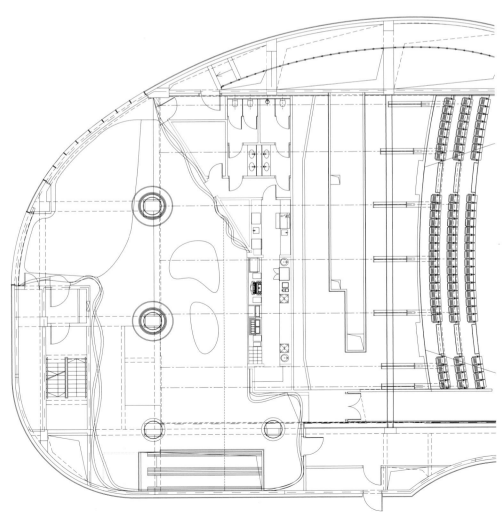

LOCATION_Kolwoon, Hong Kong, China

GH WHAMPOA CINEMA

Designer_Arthur Chan, Willie Wu, Bevin Chen
Design Company_DPWT Design Ltd.
Area_5,000m²
Material_spray paint glass, artificial stone, gypsum board
Photographer_Diamond Chan

The concept is to deliver a neat and simplicity design for the whole public lobby. Reflective glass and light trough are extensively used in the public area. The white floating corian counter, colonnade display wall accentuate a stylish ambience for the trailer/movie poster to highlight the ongoing movie. The linear articulation in lobby provides framework to display, to conceal lighting and to form a colonnade to guide patrons to the 4 cinemas. The central foyer is highlighted as a more playful area with beans shaped cushion of versatile colors scattering for family and children to take seats. This is next to the candy bar which provides a very convenient place for family to take a snack before entering the movie houses. Different sizes of white tile are used in the restrooms to give a versatile but consistent design language throughout the whole cinema.

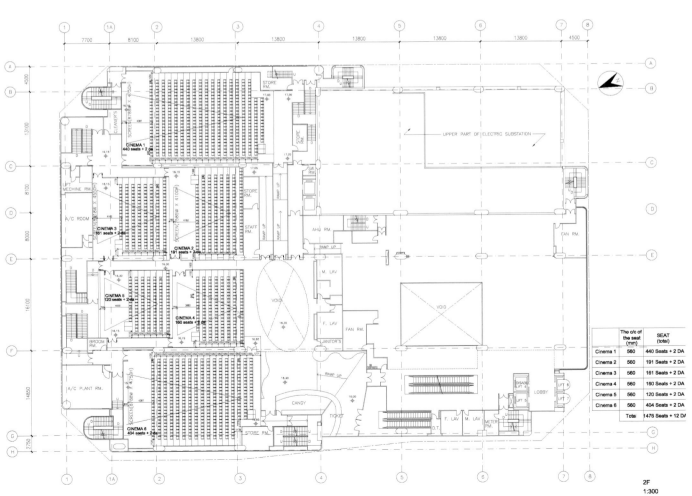

2F
1:300

整个公共大厅的设计理念是要呈现出整洁、朴素的效果。反光玻璃和灯光槽被广泛地用在公共区。白色流动型的人造石柜台和柱廊展示墙凸出了即将上映的电影预告和海报的时尚感。门厅处线性连接的框架起到展示、遮蔽舞台照明、引领顾客通往四个影厅的作用。中心大厅是一个更加有趣的区域,有很多颜色丰富的豆形坐垫可供家长和孩子选择使用。旁边就是糖果屋,供客人在观影前吃些点心,非常便利。虽然休息室内具有多种用途的白瓷砖大小不一,但整个影院在建筑设计风格上却十分统一。

LOCATION_22 Grunwaldzki Sq, Wrocław, Poland

MULTIKINO PASAŻ GRUNWALDZKI

Design Company_ Robert Majkut Design
Area_ 1,197m²
Photographer_ Szymon Polański

This project of a cinema facility in Pasaż Grunwaldzki in Wrocław is a continuation of the cooperation with Multikino company.

The entire design is focused on creating two distinct yet not separate zones: first, where the box-offices are located; and second with the main hallway leading to all of the auditoriums. This futuristic design with its science-fiction-like theme combines the two zones by using modern forms and shapes. Smooth and soft lines bring further unity to this project.

The main form for the entire design is an ellipse. The space is finished with pieces of furniture that are customer friendly, relaxing, and soft-shaped. Two broad ellipses, hanging from the ceiling and overlapping the stairway, create a central space from where the entire floor plan of the theatre's foyer begins.

The entire project is kept within a dark color scheme with walls and floor carpets that create a background for the stainless steel ceiling structures, elliptical shapes around bars and poles. Pink sofas, multimedia carriers, and recessed lights give a final touch. Additionally, recessed dark blue ceiling lighting and holographic films are used in the hallways, completing the look where light qualities are being carefully used to create an atmosphere that is desired by the movie theater customer.

The design of this cinema was a real challenge – completed on a very tight time schedule in order to celebrate the opening on the same day as the entire mall. That is why this design is open to changes needed for the most effective functioning.

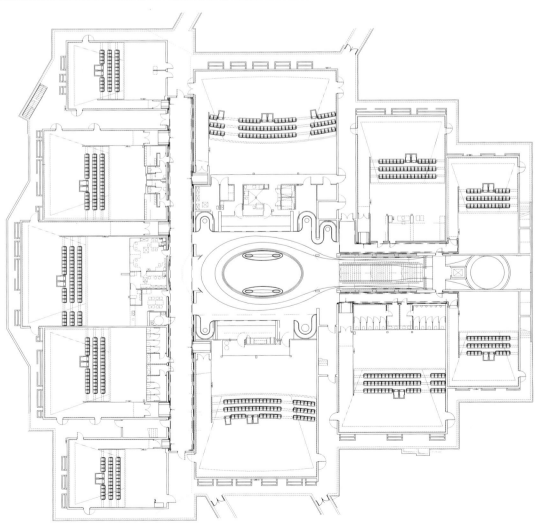

这个项目设计的影院坐落在 Wrocław 的 Pasaż Grunwaldzki 购物中心，是与 Multikino 公司再次合作的成果。

整个设计着眼于打造两个迥然不同却又相互联系的空间：首先要考虑的问题是，售票处要安排在哪里；其次是连通所有观众厅的主通道要怎么设计。这个未来派艺术风格的设计带着科幻性主题，通过现代形式和造型的使用将两个空间联系在一起。柔和的线条更加深了这个项目设计的一体化形象。

影院整体设计呈椭圆形。内部空间由客户所喜爱的、轻松自然且造型柔和的各种家具装饰而成。两个宽阔的椭圆形悬于天花板上，与楼梯相叠，形成了剧院门厅的中心区域。

整个设计中的墙壁和地毯都以暗色为主调，成为了不锈钢天花板、酒吧和圆柱周围椭圆形构造的背景。粉色的沙发、多媒体的行李架以及嵌灯起到了画龙点睛的作用。此外，为了营造出观影者期待的灯光效果，影院谨慎使用灯光，通道里天花板深蓝色的嵌入式灯光和全息胶片更加完善了这一视觉效果。

这家影院的设计是个真正的挑战——为了与整个购物广场在同一天开业庆祝，这项工程必须在极其紧迫的时间里完工。因此，影院所做的所有改变都是最高效的。

LOCATION_Mumbai, India

PVR CINEMAS PHOENIX MILLS

Design Company_ERA Architects, Mumbai and Jestico+Whiles, Prague
Area_7.897m²
Photographer_Mr. Girish Patil

One of the most premium multiplex situated in an upscale area of South Mumbai in India, PVR cinema at Phoenix Mills boasts of state-of-the-art interior design and is one of the most recognized cinema projects of Mumbai. It has seven screens in total, four are on the lower level and three on upper level, both lobbies connected with a unique ribbon-like transition element enveloping escalators. Overall lobby design plays with geometric forms, angular as well as circular blending them with play of light and color. Part of lobby ceiling is designed to resemble droplets of water, some transforming into lights, creating intrigue while the rest of it has bold angular forms studded with LED lights and a dark blue color palate. The bold forms envelopes the concession counters and color transits to black and yellow combination making it an inviting space for food and drinks. Another striking element is the "3D wall", which gives an effect of a video library stacked with movies and music. Also taking cues from cinematic elements are the drapes symbolizing show-reels of a movie. The seating niches are carved out in circular forms creating enclosures, with red hues and dim lights. Auditoriums seat around 2,000 seats and have a purple and a black color palate.

The overall design strikes a balance of architectural form, innovative use of material and lighting concepts making it one of a kind cinema design.

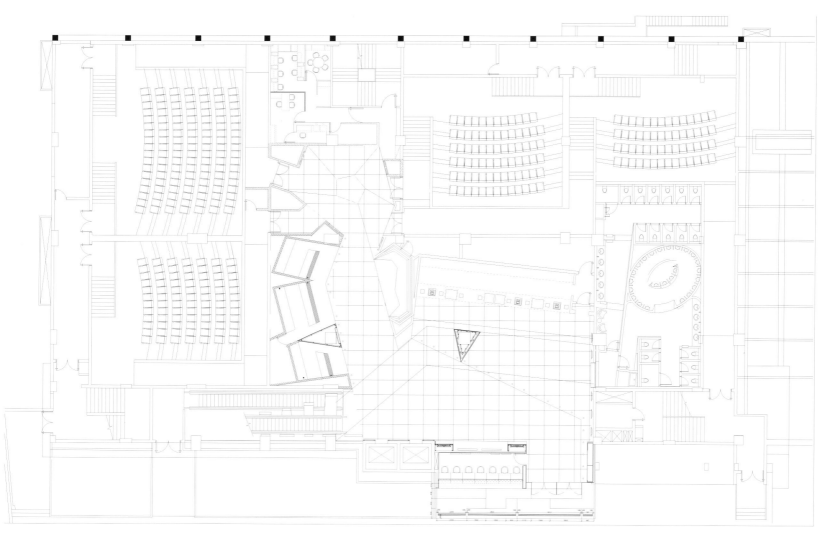

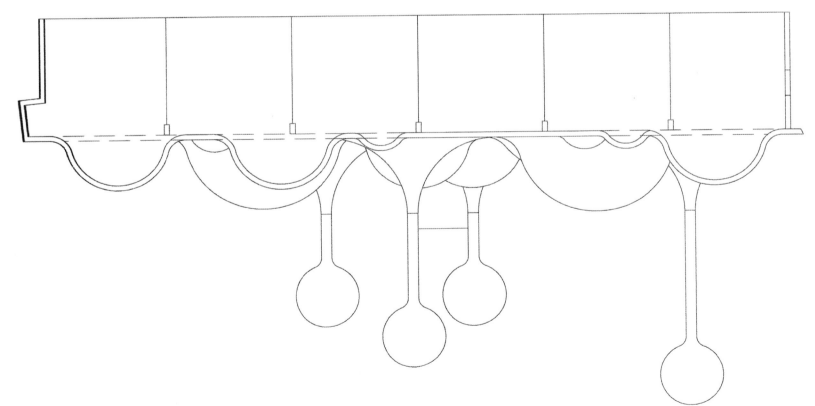

Phoenix Mills 的 PVR 影院坐落于印度孟买南部的高档消费区，是该区域内高级多厅影院之一，拥有印度最先进的室内设计，是孟买最被认可的项目之一。影院共有七块屏幕，其中四块安装在低处，三块在高处，两个大厅以独特的包围自动扶梯的带状元素相连接。大厅整体设计通过灯光和色彩的运用将几何图形、混合棱形以及圆形融合在一起。部分大厅天花被设计成水滴状，有的则改装为灯光，营造出一种神秘气氛；而其他部分的天花板则装饰以 LED 灯光和暗蓝色色调。轮廓分明的区域是接待专柜，黑黄相融的色彩代表了令人胃口大开的餐饮区。另一个重要元素就是"3D 墙壁"，呈现了一种堆满了电影和音乐的影像图书馆的效果。设计还从电影元素中得到了启发，即帷幕象征了电影的演出卷轴。安有座椅的圆形区域形成一个封闭空间，以红色及朦胧的灯光做装饰。观众席容纳了大约 2000 个座椅，座椅有紫色和黑色两种颜色。

整个设计打破了传统建筑模式的平衡，对材料、灯光的创新型应用使其成为独一无二的影院设计。

LOCATION_Hefei, China

MBOX CINEMA

Designer_James Law
Photographer_Chester Ong

The rapid change of films in the film industry, the cinema is where the audience experiences different journeys. The circulation of this energy is the design concept of this project; our aim is to create a landmark full of dynamics. Our interior design is inspired by actions and speed to bring up the tension in movies, creating an unordinary experience for the audiences.

Exterior Design: The three-dimensional logo is a multi-faceted LED lights sculpture, extending through the glass and out of the exterior next to the entrance. It continuously plays trailers of new movies, giving the audience a brand new impression of a contemporary cinema layout design. It is a way of promoting their cinema, to let everyone know about MBox cinema. The MBox logo is projected onto the building using laser technologies just like projecting a film, creating vibrant images and an extraordinary impression.

Lobby Design: The entire space of the building is designed with a contemporary approach to bring up the vibrant images from movies to give customers an extraordinary experience. There are four irregular independent shaped ticket counters which lights up from below has a touch screen surface on the table which is designed to break a tradition ticket counter. This design is also aimed to bring the distance of the worker and the customer closer, to obtain better services.

The three-dimensional panels are constructed by triangle and multimedia. This includes three-dimensional triangular LCD screens, merging with movie trailers where customers can watch different trailers before purchasing their tickets. The interior design also gives the effect of jumping into virtual spaces. LED electronic posters are used as it is environmentally friendly and to give a more advanced feeling. The reflective wall surfaces create interesting patterns where customers could be entertained while waiting for someone. An electronic message board is also installed to provide information to customers. In addition, the waiting area fitted with black long benches is extended from the walls of the three-dimensional effect; this makes the interior space more connected and innovative.

Cinema Design: The inspiration of the interior design is based on the concept of sound waves, moving out from the screen into the audience. Lights are installed into the walls, hidden in the blue atmosphere. The Light gradually changes and the speakers are hidden to reduce visual obstructions. The seats are specially arranged so that the audience could watch the film at the best angle. Gray carpets are used to give customers a comfortable environment and it helps reduce the noise generated. We believe that this theatre will inspire a special, unique and new experience to demand audiences.

　　随着电影业的飞速发展，影院成为了观众体验不同旅程的场所。这个项目的设计理念是能源的循环使用，设计的目标是打造出一个充满动感的地标性建筑。室内设计的灵感来源于动作与速度，通过增强影片的紧张气氛给观众带去非同一般的体验。

　　外部设计： 影院商标的 3D 效果是用 LED 灯光多面塑造而成的，通过玻璃一直延伸到入口附近。影院循环播放最新影片的预告片，让观众对现代影院的布局设计有个新的印象。这也是提升影院的一个方法，让每个人都了解 MBox 影院。设计师采用镭射技术，像播放一部影片一样在影院建筑上放映 MBox 影院的标识，营造出活力四射的形象，留给人深刻的印象。

　　门厅设计： 建筑的整体空间采用现代设计手法，加强了影片的画面效果，力求带给观众超凡的体验。四个独立售票柜台打破了传统售票柜台的构造，呈现出不规则的形状，柜台的灯光自下而上，并且每个柜台台面都配有一个触摸屏。这个设计同时也旨在拉近工作人员与消费者间的距离，以便提供更好的服务。

　　3D 的墙壁呈现出三角形、多媒体效果。墙壁上有三角形的 3D LCD 显示屏，屏幕上播放着电影预告片，观众可以在购买电影票之前就从这里欣赏到不同的预告片。室内设计给人跃入虚拟空间的感觉。LED 电子海报因其绿色环保而被用于影院内，带给观众更多高级的享受。带有反射性的墙体表面呈现出有趣的图案，观众可以在等待同伴时自娱自乐。影院安装的一块电子信息板给消费者提供了必要的信息。此外，候客区从有 3D 效果的墙壁上延伸出来，内设黑色长椅，使室内空间更具整体性和创新性。

　　影院设计： 室内设计灵感源于声波，起于荧幕而后进入观众耳中。灯安装在墙壁里，隐匿在蓝色氛围中。灯光逐渐变换，扬声器被隐藏起来以保证视觉通畅。独特的座椅排列使观众能够以最佳的角度观看影片。灰色地毯给观众营造一种舒适的环境，并有助于减少噪声。我们相信这家影院能给前来的观众带来独特的、非凡的全新体验。

THEATRE

剧院

LOCATION_Oslo, Norway

OPERA HOUSE OSLO

Architect_Snøhetta AS
Area_38,500m²
Material_stone, timber, metal, glass
Photographer_Birdseyepix, Helene Binet

The Norwegian National Assembly voted in 1999 that the operahouse was to be built in Bjørvika, on the seafront at the Oslo fjord. In this way the building would be the foundation for the urban redevelopment of this area of the capital. The main auditorium is a classic horseshoe theatre built for opera and ballet. It houses approx. 1,370 visitors divided between stalls, parterre, and three balconies. Technical space occupies the area above balcony 3. The orchestra pit is highly flexible and can be adjusted in height and area with the use of three separate lifts. On each side of the stage are mobile towers which allow for adjustments in the proscenium width for ballet or opera without damaging the acoustics of the hall. Reverberation time is fine tuned using drapes along the rear walls and control rooms for sound and light are located to the back of the hall. The form of the auditorium is based on several relationships: short distance between the audience and the performers, good sight lines, and, above all, excellent acoustics. The architectural intentions for a modern auditorium with traditional, acoustic musical performance have been developed in parallel with requirements for visual intimacy and acoustic excellence. In older opera halls acoustic attenuation was often achieved by using rich decorative, sculptural elements on most surfaces. In this case the requirements have been met with a clean, carved aesthetic using a modern formalistic language. The requirement for a long reverberation time results in a room with a large volume. In this case the volume is increased by the use of a technical gallery which cantilevers out over the walls below, giving the hall a T shaped section. The main structure of the stone clad roof above is included in the volume of the hall rather than being hidden behind a false ceiling.

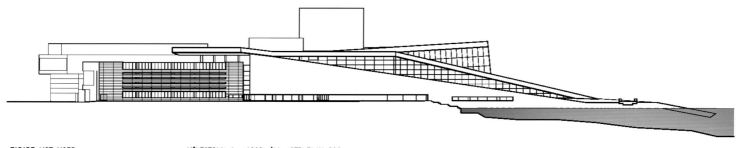

FASADE MOT NORD   MÅLESTOKK 1 : 1000 /A4 CTB PLAN 200

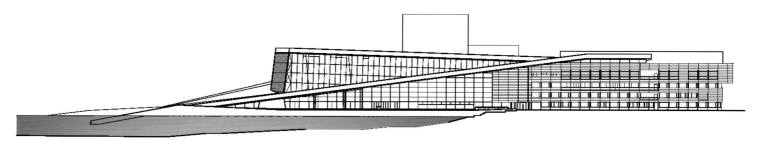

FASADE MOT SØR   MÅLESTOKK 1 : 1000 /A4 CTB PLAN-200

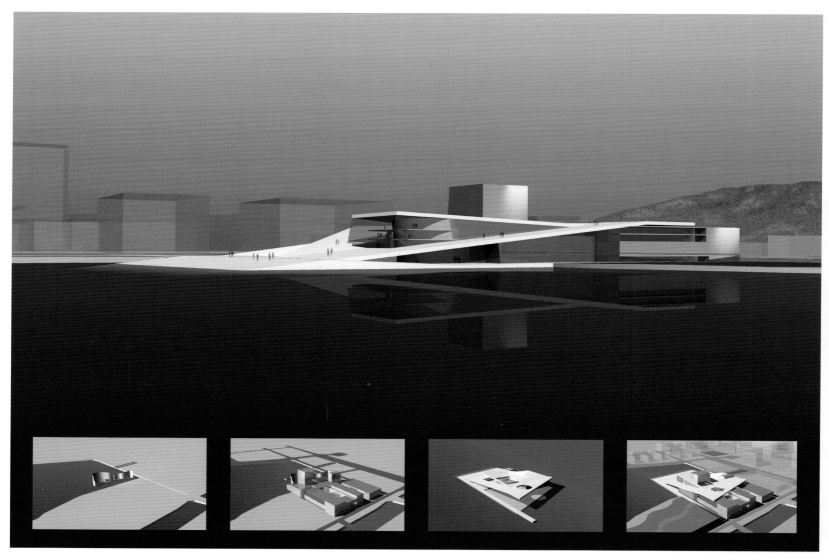

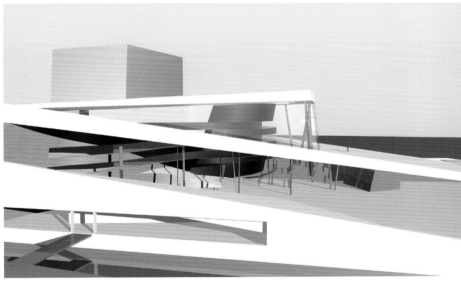
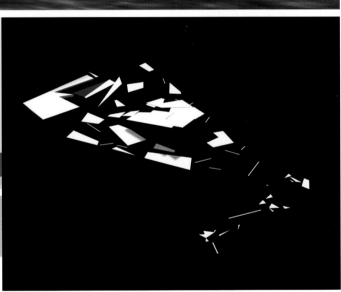

123

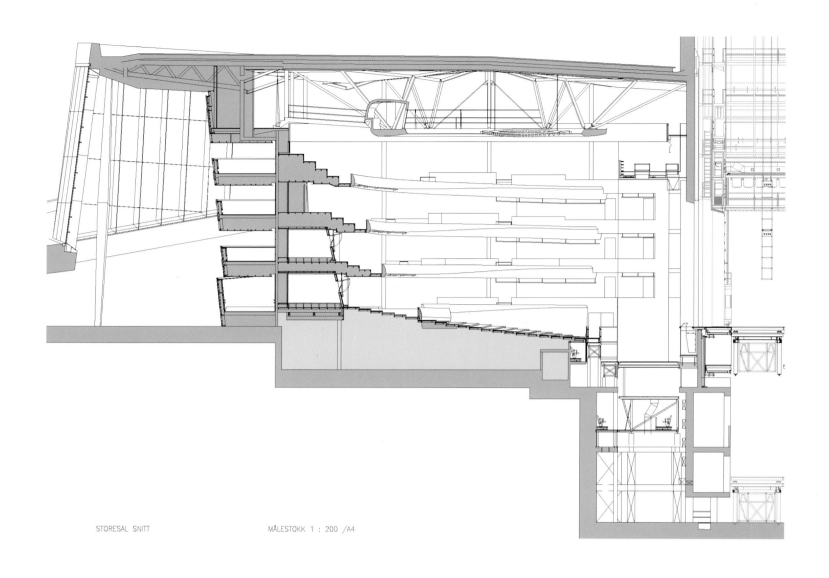

STORESAL SNITT   MÅLESTOKK 1 : 200 /A4

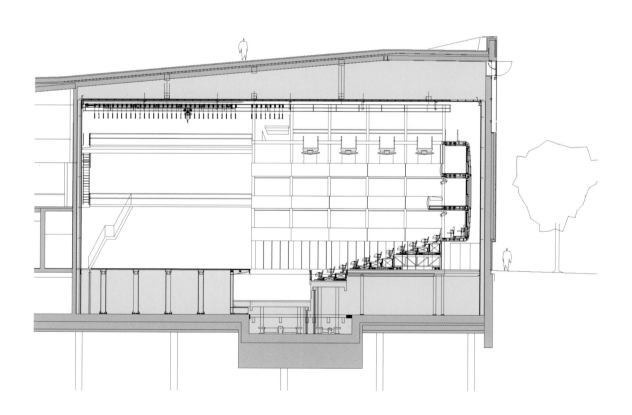

LILLESAL 1:200 A4

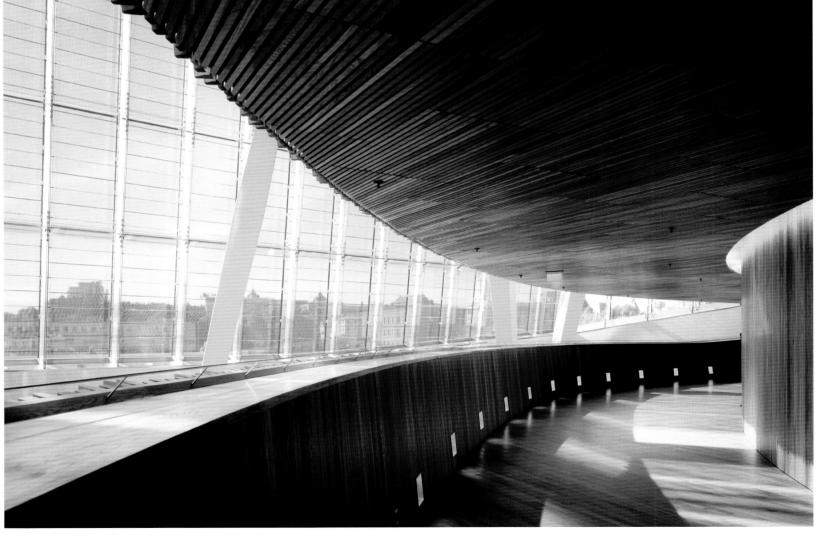

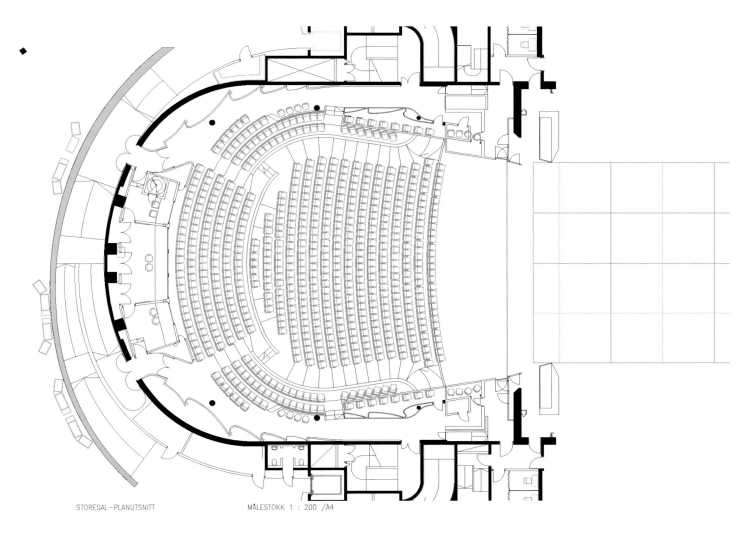
STORESAL–PLANUTSNITT   MÅLESTOKK 1 : 200 /A4

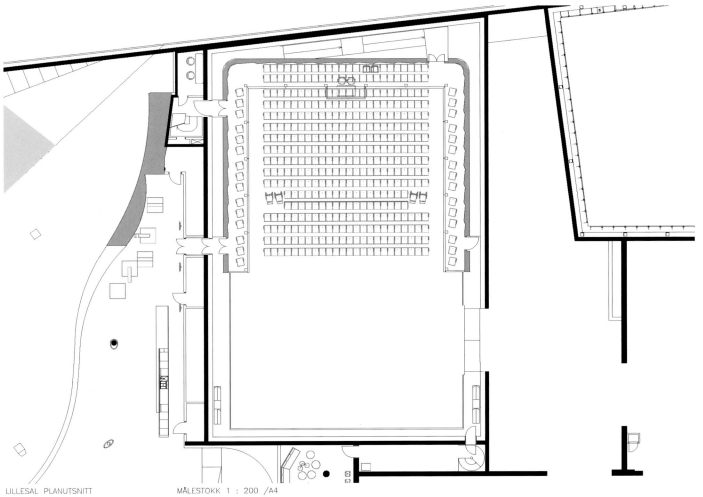
LILLESAL PLANUTSNITT   MÅLESTOKK 1 : 200 /A4

挪威国民大会1999年投票通过了在奥斯陆海湾海滨地带Bjørvika修建奥斯陆歌剧院的决议。这样的话，该剧院将成为首都地区再发展的基础。马蹄形的主剧院是专为歌剧和芭蕾舞剧设计的。大概1370个座位被分列在前排、后排和三个阳台中。第三个阳台上面是技术支持室。乐池的设计相当灵活，可通过三个独立的电梯调节高度和面积。舞台的每一侧都有可移动的塔台，这样便可根据芭蕾舞剧或歌剧来调节舞台的宽度，不至于破坏音质。混响时间通过运用后墙的布帘得到了很好的调整，声光控制室位于大厅的后面。礼堂的修建模式是根据一些关系来确定的：观众和表演者间的近距离接触，恰当的观赏距离，尤其是好的音质。该剧院的设计目的是为了打造出集传统风格和声乐表演为一体的现代礼堂，并达到视觉和听觉的最佳效果。在旧歌剧院大厅，通常是用大量的装饰和墙壁表面的雕刻来吸收声音，从而减弱声音；而此案则是运用了现代形式主义的设计语言营造出简洁的雕刻美感。室内面积足够大才能营造出长时间的混音效果。在此项目中，设计师通过巧妙的走廊设计扩大了室内空间，走廊呈悬臂式伸出墙外，与其下的墙面一起营造出了T形的大厅区域。石质屋顶的主要结构位于大厅的上空，而不是被隐藏在假天花板之后。

TAKPLAN　MÅLESTOKK 1 : 1000 /A4

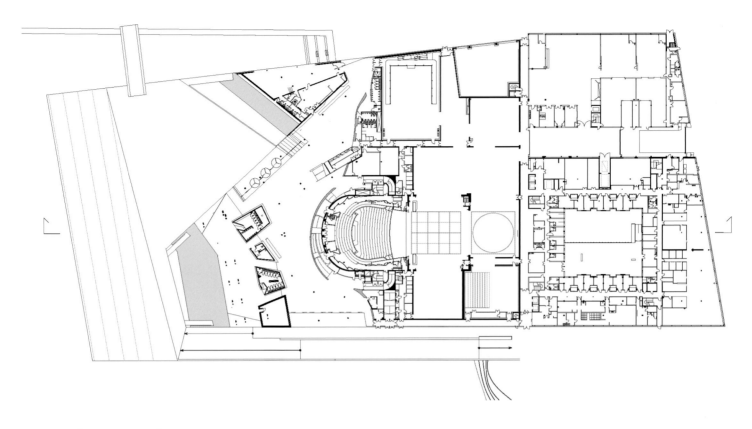

PLAN 1   MÅLESTOKK 1 : 1000 /A4

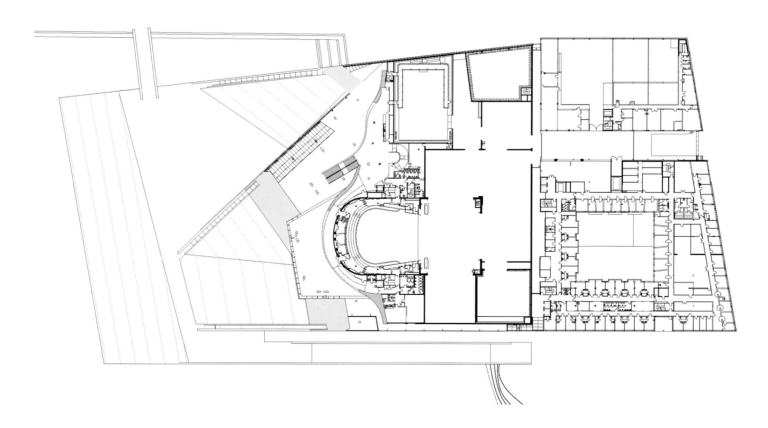

PLAN 2   MÅLESTOKK 1 : 1000 /A4

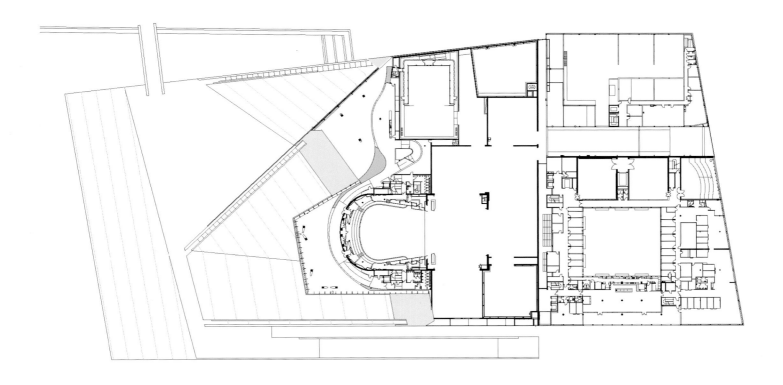

PLAN 3   MÅLESTOKK 1 : 1000 /A4

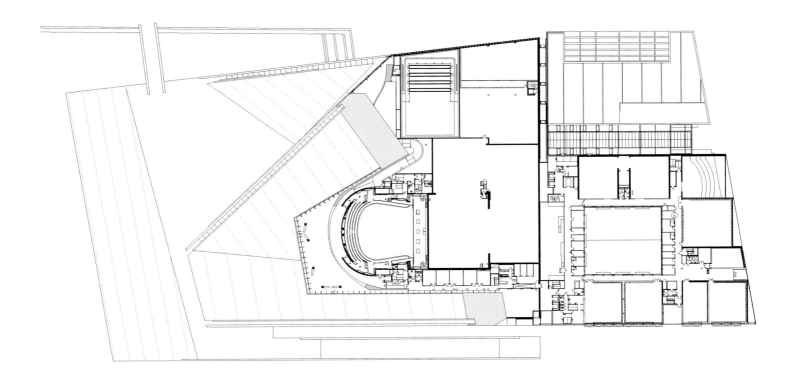

PLAN 4   MÅLESTOKK 1 : 1000 /A4

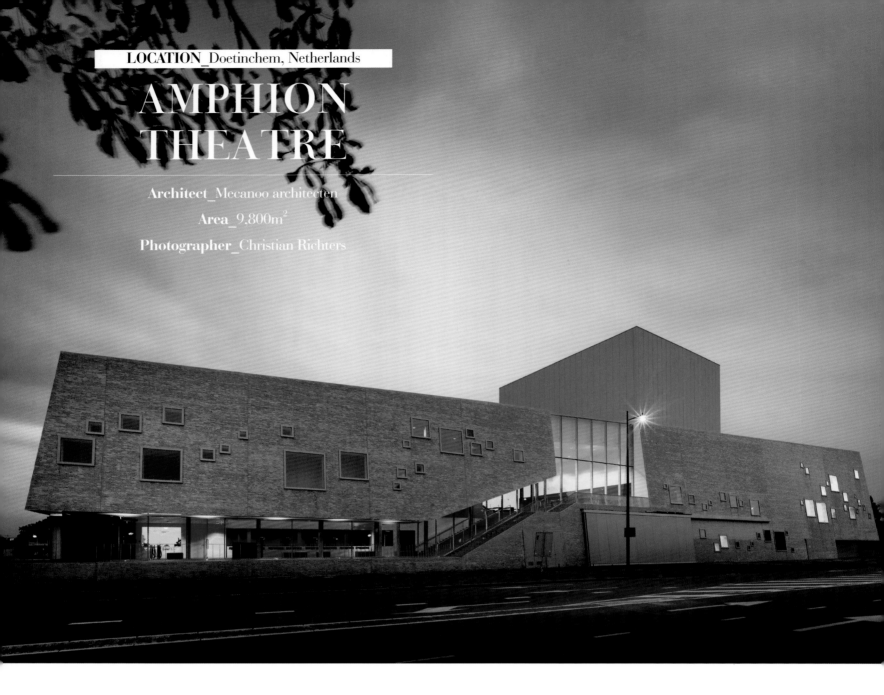

LOCATION_Doetinchem, Netherlands

AMPHION THEATRE

Architect_Mecanoo architecten

Area_9,800m²

Photographer_Christian Richters

The audience enters the theatre through the courtyard which takes the role of a red carpet. Red accented incisions mark the entrance, the theatre cafe and the green room. Once inside you venture into a world where warm red shaded tones prevail. A grand staircase leads to the theatre cafe, the central meeting place in the theatre from which the four foyers and large and small theatre halls are accessible. The beautifully detailed bar is set like a jewel in the double-height space. The window pattern in the façade creates surprising compositions, imbuing each space with its own character. The windows appear like paintings overlooking the city, sometimes serving as a standing table or bench with their deep sills. The large 860 seat auditorium is like a treasure hidden within the building. Only the fly tower on the exterior betrays the presence of the auditorium. The intimate horseshoe-shaped room in various shades of red is reminiscent of the theatres of yesteryear. Three balconies encircle the theatre, seating the audience close to the stage. Visitors have a perfect view from every seat onto the imposing 36 x 19 meter stage floor which is suitable for all large-scale theatrical productions. The small 300 seat hall can be used as a Black Box and is almost an exact copy of the popular small theatre hall of the old Amphion. The U-shaped layout of movable stands ensures direct contact between actors and spectators. By moving the stands together, the venue is transformed into a concert hall or a ballroom.

In the Amphion theatre, performances are changed at a relatively high speed. To facilitate this, efficient loading and unloading is required. Mecanoo provided the loading and unloading area with enough space for three 18-meter long trucks, centred between the stages of the two theatre halls located on the same floor level. This space can also be used to temporarily house stage sets or to exchange sets between the large and small theatre halls. The main hall is equipped with fully automatic features in the 29 meter high fly tower. Uniquely, the large and small theatre halls have windows, so that the technical staff has daylight in their workplace.

• Siteplan

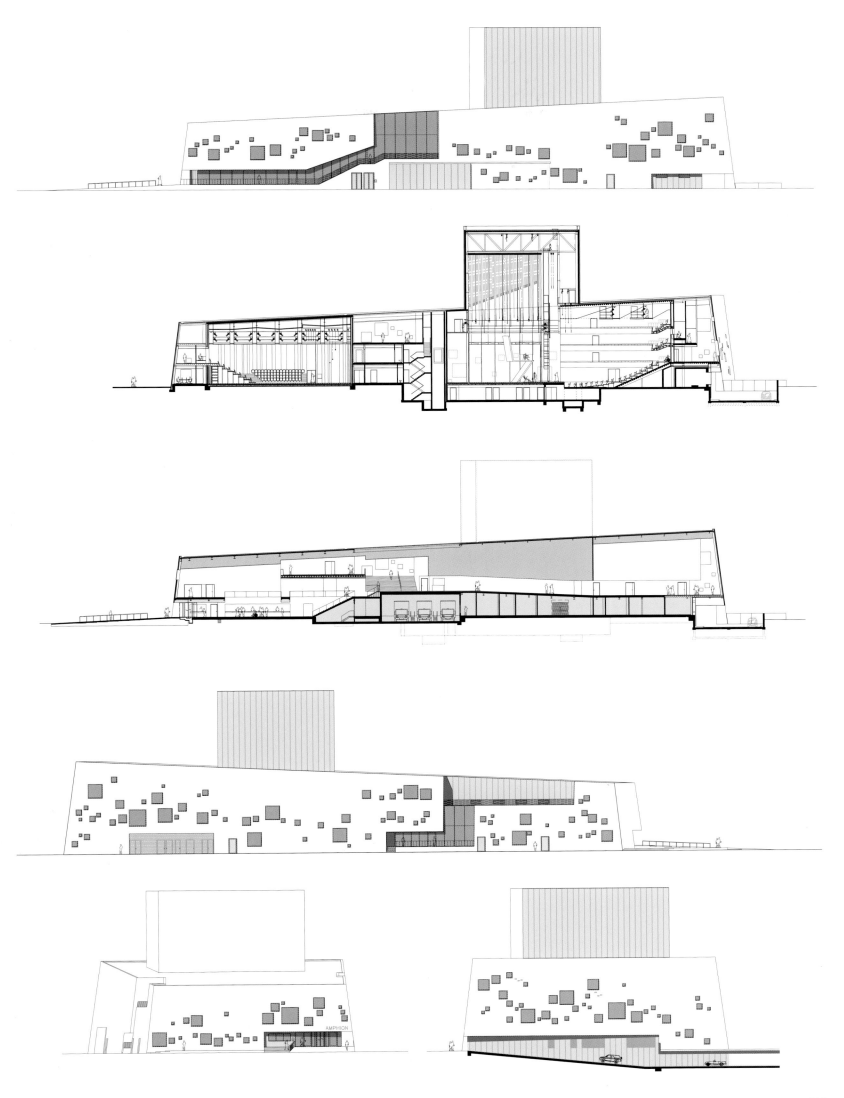

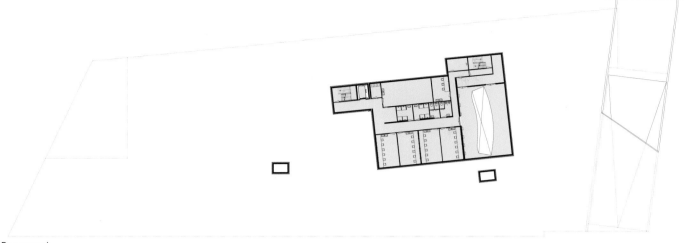
Basement

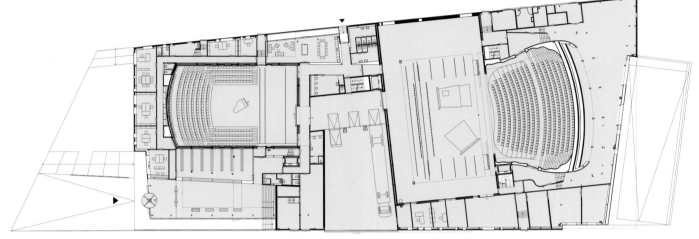
Ground Floor

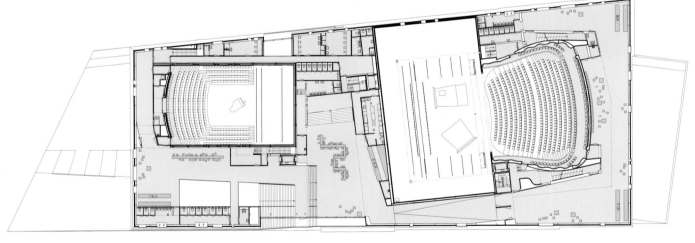
First Floor

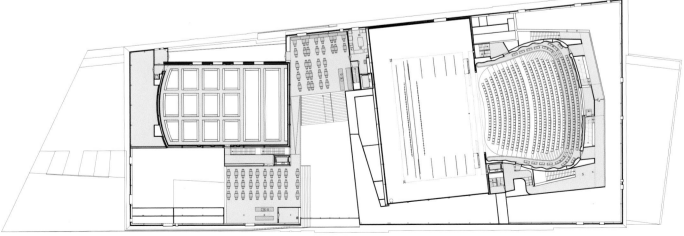
Second Floor

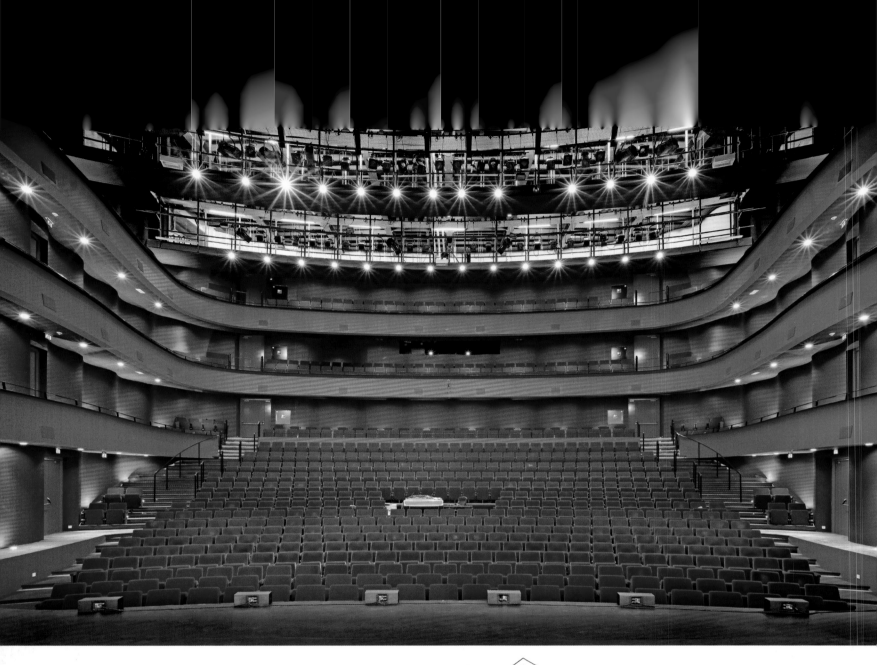
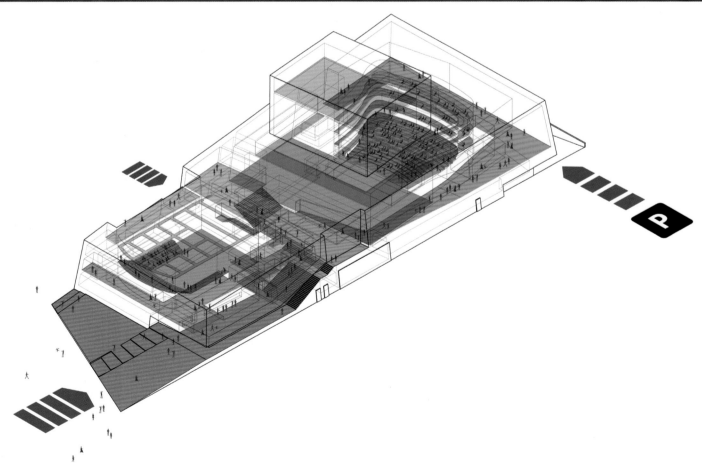

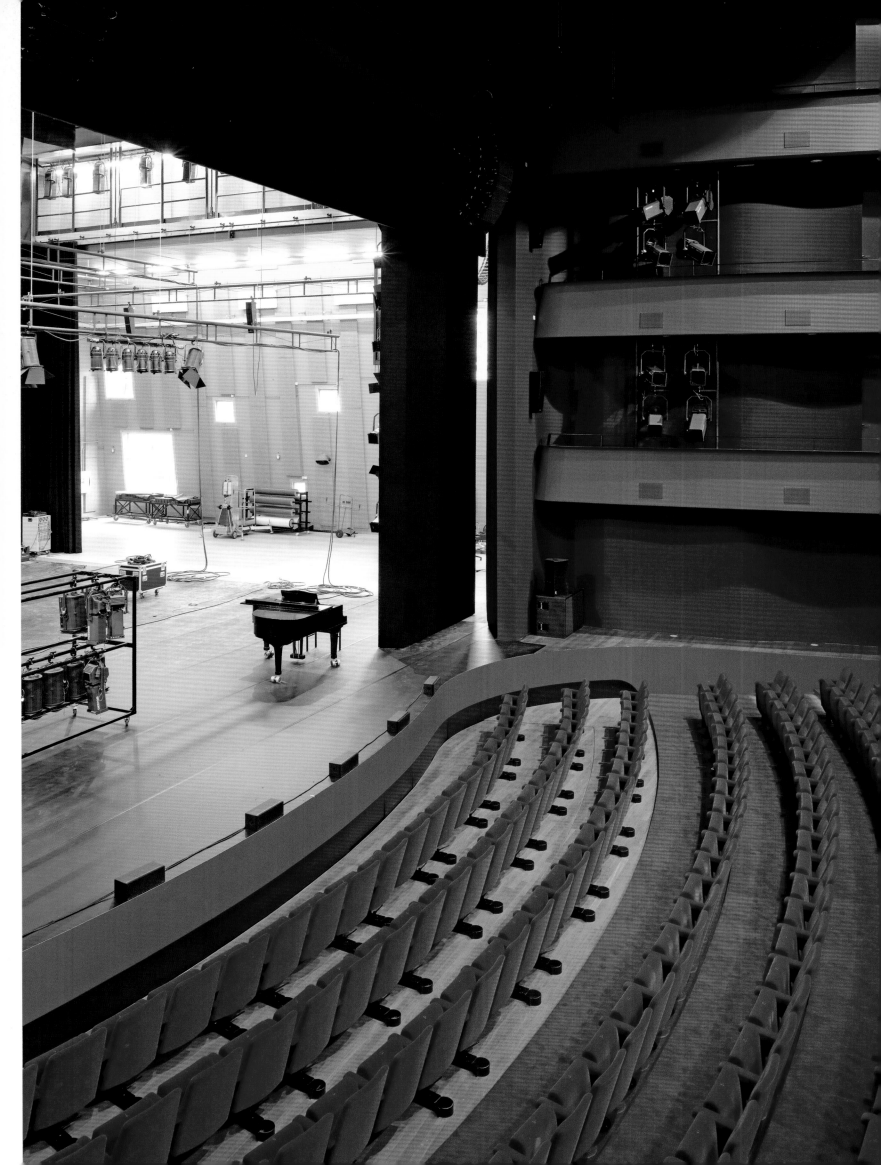

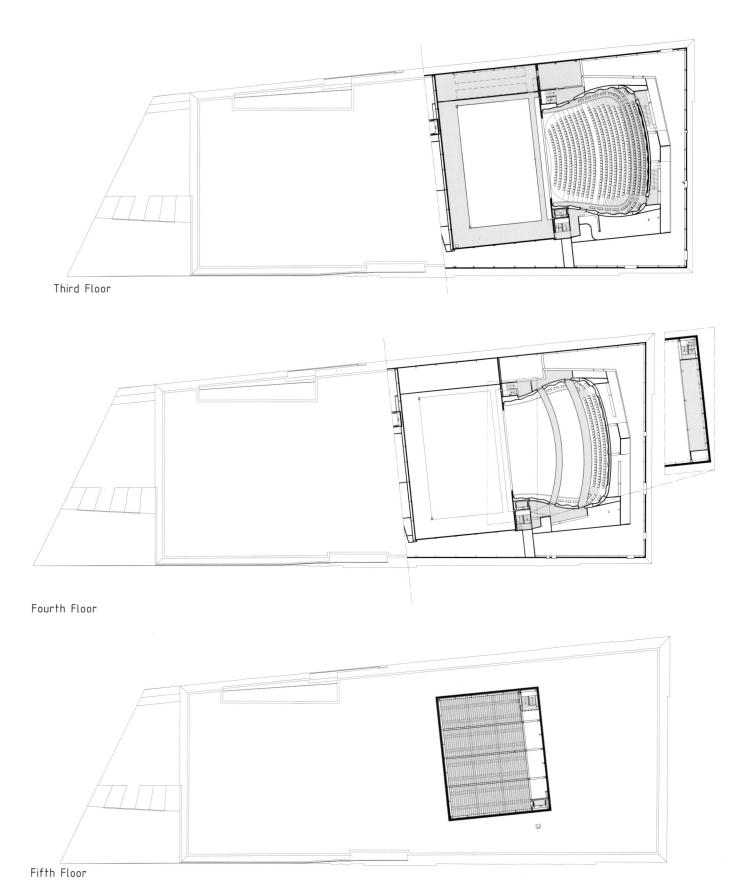

Third Floor

Fourth Floor

Fifth Floor

　　观众通过扮演着红地毯角色的庭院进入剧场。红色醒目的雕刻标志着入口、剧院咖啡厅和演员休息室的方向。一旦进入剧院，你就仿佛进入了一个暖色调的红色世界。一段气派的楼梯通往剧院咖啡厅，这里是剧院的中央广场，从这里可到达四个门厅和其他大大小小的剧场。精致的酒吧如同镶嵌在这个双层空间里的一颗明珠。外立面的窗口格局使得整个建筑呈现出一种令人惊讶的独特构造，使每个空间都独具特色。窗户看上去就像油画一般，俯瞰全城，窗台那么深，有时也能起到常设桌椅的作用。可容纳 860 个席位的大礼堂，如同建筑物内隐藏的宝藏一般。只有从外部的台塔才能看出礼堂的存在。亲切的马蹄形空间采用了深浅不一的红色调，让人想起昔日的剧院。三个阳台包围着剧场，使观众的座位靠近舞台。舞台长 36 米，宽 19 米，适用于所有的大型舞台演出，无论观众在哪个座位都能获得绝佳的观赏角度。可容纳 300 个席位的小厅可用作黑盒子，几乎与老 Amphion 剧院里受欢迎的小演出厅一模一样。可移动看台呈 U 形布局，确保演员和观众之间可近距离接触。将看台挪到一起，还能把场地转化为音乐厅或舞厅。

　　在 Amphion 剧院，演出变化是相当频繁的。因此，必须保证较高的上场和谢幕效率。在两个舞台剧场中间的同一水平高度上，Mecanoo 设计师事务所设计了足够三辆 18 米长的卡车停放的转换空间，位于同一楼层的两个剧院大厅舞台的中央。这个转换空间可用于临时存放舞台布景，也可以实现大小剧场之间的背景转换。主要大厅配备有 29 米高的全自动舞台塔。与众不同的是，大小剧场大厅都设有一些窗口，让技术人员在工作场所也可享受到自然光线。

LOCATION_ George Street, Corby, Northants NN17 1QB, UK

CORBY CUBE

Architect_Hawkins\Brown

Area_7,700m²

Photographer_Tim Crocker, Hufton+Crow

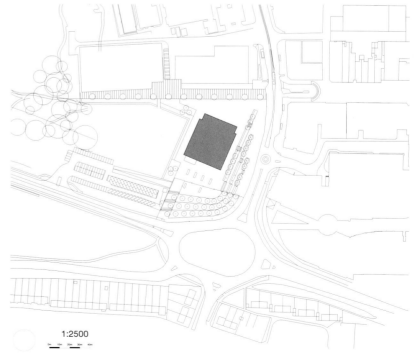

1:2500

Hawkins\Brown's striking glazed Corby Cube, the focal point of a major regeneration programme to revitalize the former steel making town in Northants, UK has completed and is now open to the public.

Won in international competition in 2004, the Corby Cube combines a diverse programme of civic, and arts uses into a new model for civic architecture. The resultant 7,700 sq m building features a mix of facilities including: A 450-seat theatre with front of house foyers and bars; A multifunctional studio space; A ground floor Café.

Designed in collaboration with theatre designers Charcoalblue, the aim of the theatre was to create an intimate space where members of the audience could experience a variety of different acts at close quarters. The brief for the theatre evolved throughout the consultation process from the original vision of a formal fly tower theatre with proscenium arch, wings, raked stalls and traditional raised stage, to the realized design of a flexible modern venue that incorporates all the qualities of a traditional variety theatre. The theatre incorporates a range of innovative technology, including: a curved retractable seating system, which is the first of its kind in the UK; a seating elevator and wagon that can be raised and lowered to accommodate a range of formats; a movable proscenium combine to provide maximum flexibility to accommodate a diverse programme of events.

Influenced by several traditional theatres such as London's Royal Court and Old Vic, the interior of the theatre is imbued with a sense of quality through the selection of a rich palette of walnut and plush fabrics, including the specially commissioned lining of the balcony fronts and cornice. The seating arrangement is deliberately intimate to encourage interaction of the audience in each performance, acknowledging that theatre is a participatory rather than passive experience.

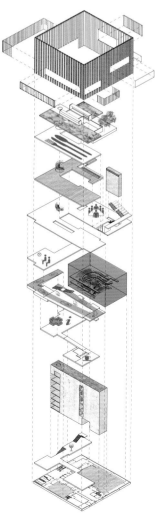

- South Elevation 500

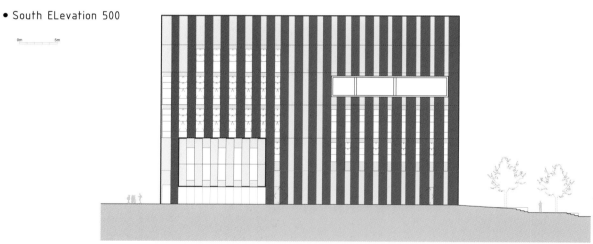

- West Elevation 500

1:500

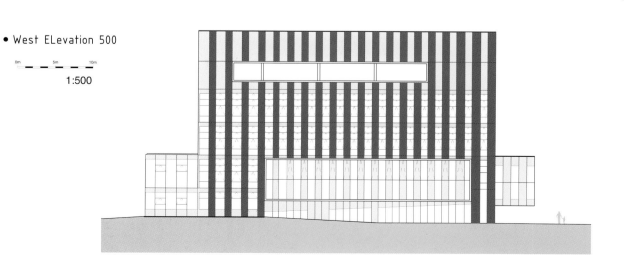

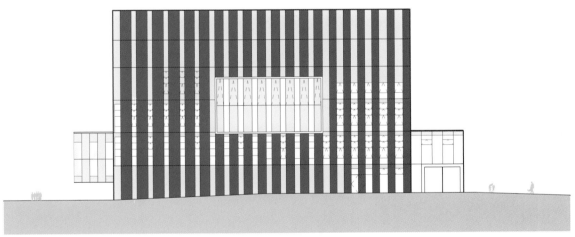

• East ELevation 500

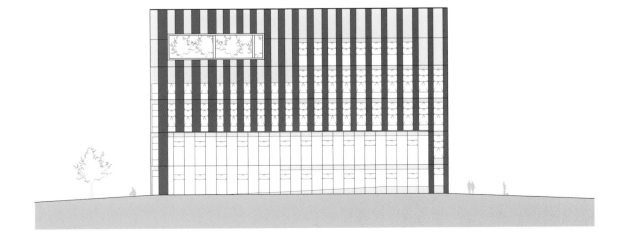

• North ELevation 500

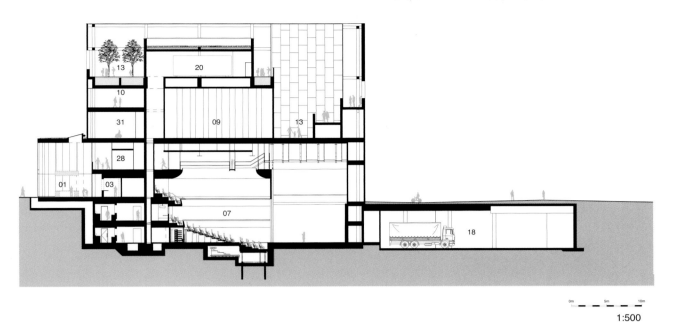
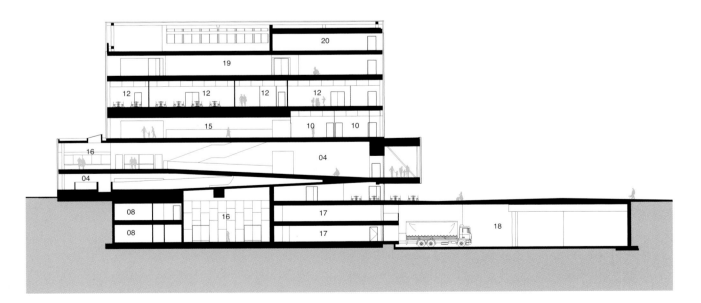

01 - Reception
02 - Foyer
03 - Box Office
04 - Library
05 - Bistro
06 - Kitchen
07 - Theatre
08 - WCs
09 - Council Chamber
10 - Meeting Room
11 - Staff Lounge
12 - Office
13 - Roof Terrace
14 - Store
15 - Circulation
16 - Studio
17 - Dressing Rooms
18 - Undercroft
19 - Restaurant
20 - Plant
21 - Bar
22 - Security Room
23 - Theatre Get-in
24 - Wardrobe
25 - Control Room
26 - Green Room
27 - One Step Stop
28 - Interview Room
29 - Technical Gallery
30 - Public Gallery
31 - Members' Lounge

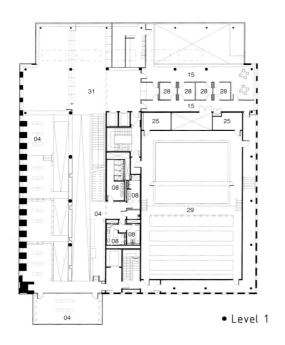

• Level 1

1:500

1.	Reception	17.	Dressing Rooms
2.	Foyer	18.	Undercroft
3.	Box Office	19.	Restaurant
4.	Library	20.	Plant
5.	Bistro	21.	Bar
6.	Kitchen	22.	Security Room
7.	Theatre	23.	Theatre Get-In
8.	Wcs	24.	Wardrobe
9.	Council Chamber	25.	Control Room
10.	Meeting Room	26.	Green Room
11.	Staff Lounge	27.	One Step Stop
12.	Office	28.	Interview Room
13.	Roof Terrace	29.	Technical Gallery
14.	Store	30.	Public Gallery
15.	Circulation	31.	Members' Lounge
16.	Studio		

1.	Reception	9.	Council Chamber	17.	Dressing Rooms	25.	Control Room
2.	Foyer	10.	Meeting Room	18.	Undercroft	26.	Green Room
3.	Box Office	11.	Staff Lounge	19.	Restaurant	27.	One Step Stop
4.	Library	12.	Office	20.	Plant	28.	Interview Room
5.	Bistro	13.	Roof Terrace	21.	Bar	29.	Technical Gallery
6.	Kitchen	14.	Store	22.	Security Room	30.	Public Gallery
7.	Theatre	15.	Circulation	23.	Theatre Get-In	31.	Members' Lounge
8.	Wcs	16.	Studio	24.	Wardrobe		

• Lower Level 1

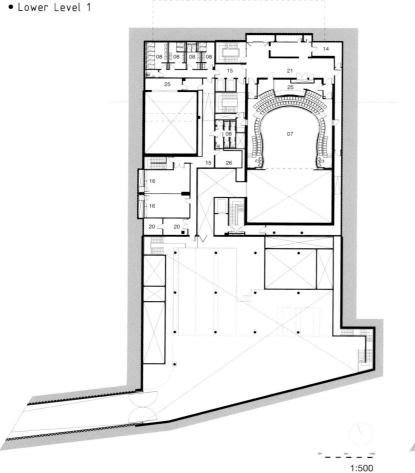

1:500

• Lower Level 2

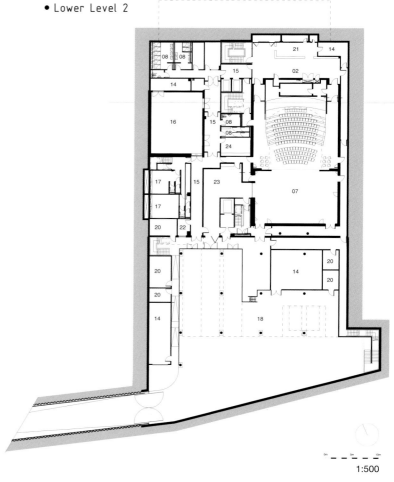

1:500

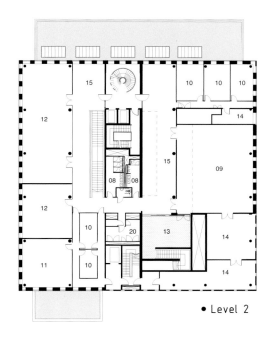

• Level 2

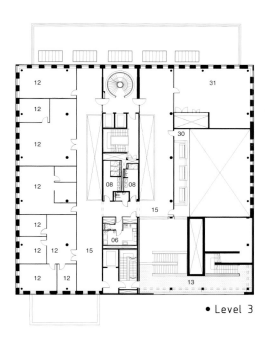

• Level 3

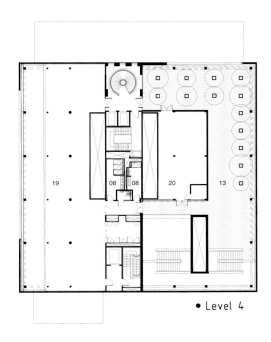

• Level 4

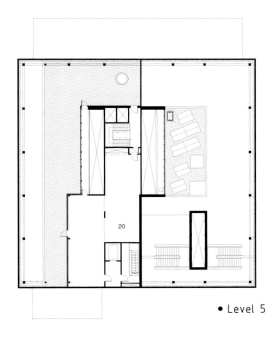

• Level 5

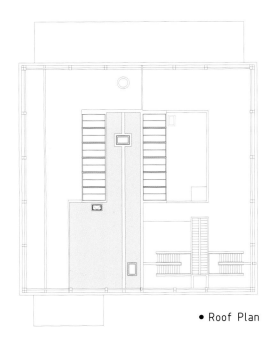

• Roof Plan

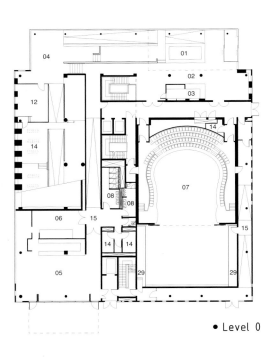

• Level 0

155

　　由英国 Hawkins\Brown 公司设计的引人注目的、装有镜面玻璃的"Corby Cube"大楼是英国北安普敦郡原钢铁小镇复兴工程的一部分，现已完工并对公众开放。

　　赢得了 2004 年国际竞赛大奖的 Corby Cube 大楼结合了城市和艺术的多样性设计规划理念，成为了新的城市建筑的典范。这个占地 7700 平方米的建筑包括一系列设施：一个有 450 个座位、门厅和酒吧的剧院；一个多功能工作室和一家坐落在一层的咖啡厅。

　　与影剧院设计师 Charcoalblue 合作设计的该剧院，旨在为观众营造一个可以近距离观看不同表演的温馨空间。剧院的发展离不开协商的过程，从最初的带有舞台拱形墙、侧翼、倾斜的分隔栏和传统的立式舞台的正规台塔式剧院发展成融合了传统的剧院特质且具有灵活性的现代剧场。剧院采用了一系列创新科技设备：弧形的可伸缩的座位系统（英国首次引进该系统的剧院）；座椅电梯和车辆可根据不同的演出阵容做出相应的升降调节；可移动的舞台能够根据不同的演出规划做出最大限度的灵活调节。

　　因为受到像伦敦 Royal Court 和 Old Vic 这样的传统剧院设计的影响，此剧院的内部设计给人一种质的感觉。空间选用胡桃木的深色基调，大面积应用奢华的毛绒面料，此种色调还体现在阳台及檐口的设计上。精心安排的座位加强了观众与演员在每场表演之间的互动，强调了戏剧是一种分享的过程而非被动的体验。

LOCATION_ Canterbury, Kent, UK

MARLOWE THEATRE

Architect_Keith Williams Architects
Area_ 4,850m²
Photographer_Hélène Binet, Tim Stubbins

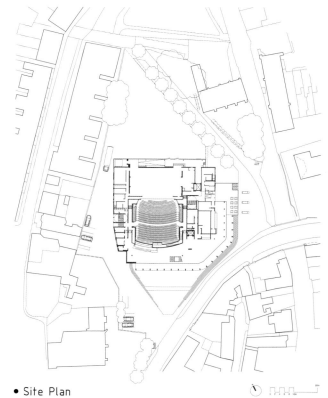

• Site Plan

Named after Christopher Marlowe, the city's famous Tudor playwright, this iconic modern theatre in the heart of the historic city of Canterbury has been designed by the award-winning London-based Keith Williams Architects. Standing on the banks of the River Stour, nearby Canterbury Cathedral's UNESCO World Heritage Site, the new Marlowe Theatre makes a bold statement on the Canterbury skyline. The project has been constructed on the site and adjoining lands of the old Marlowe Theatre, itself a converted 1930s cinema, which was demolished to make way for the new project. Williams' competition winning project for the 4,850 sqm new Marlowe is, in formal terms, a complex pavilion. It sets up a dynamic relationship with its viewers, giving different architectural and urban emphasis depending from where in the city it is viewed. At street level, its architecture is ordered by an 8m high colonnaded loggia in white cast Dolomite stone, which forms a portal to the multi-level glazed foyer, and sets up a civic elevation to the Friars, an important historic street within the city. The foyer connects all the major internal spaces to the riverside terraces and pathways, and is seen as a crystal ribbon by day transforming into a blade of light by night. New views of the roof tops of the historic city and its cathedral open up from the main stairs and upper levels.

The flytower is clad in a stainless steel mesh skin held 600mm off a weathering skin of silver anodised aluminum panels, causing its form to dematerialize and its surfaces to shimmer and sheen whilst subtly reflecting the changing hues of the daytime sky and sunset. Internally, the double height foyer and feature staircase leads to the main auditorium set over three levels and lined with black American walnut. The auditorium, lined in black American walnut, seats an audience of 1,200 in fiery red/orange leather seats. A second smaller performance space, the Marlowe Studio, a flexible format studio theatre gives seating 150. The Marlowe Studio is placed 6m above the foyer allowing its spaces to flow uninterrupted toward the riversides terraces and bankside, whilst also giving views toward the Cathedral. The public areas including bars and cafes, a new riverside walk and piazza are open throughout the day where visitors can enjoy a year round programme of daytime activities and exhibitions.

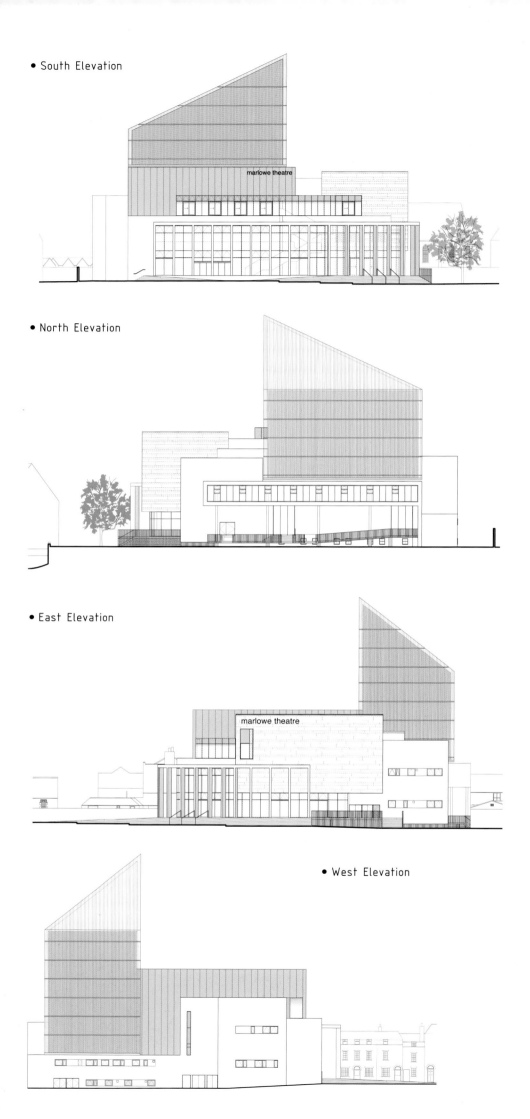

- South Elevation
- North Elevation
- East Elevation
- West Elevation

Marlowe 剧院是以都铎王朝时期著名的剧作家 Christopher Marlowe 命名的，这个具有标志意义的现代剧院坐落于历史古城坎特伯雷的中心，由总部位于伦敦的一流设计公司 Keith Williams 设计师事务所设计。坐落于斯陶尔河畔，临近被联合国教科文组织列为世界文化遗产的坎特伯雷大教堂，新 Marlowe 剧院成为坎特伯雷地平线上浓墨重彩的一笔。新 Marlowe 剧院在其原址上动工，占地面积更广，原建于 20 世纪 30 年代的电影院，为了新剧院的建设，已被拆除。关于新 Marlowe 剧院的设计方案竞争中，Keith Williams 设计师事务所的设计胜出，按照其规划设计，这个占地 4850 平方米的新 Marlowe 剧院将是一个综合展馆。建筑与观众们建立起动态的联系，无论从城市的哪个方向望去，都能展现出城市和建筑的不同特色。在街道层面上，建筑物是由八根白色云岩质的柱子支撑的，形成了进入多层釉面门厅的入口，也建起了一个通往城内历史性街道 Friars 的平台。门厅连接起室内所有的功能空间，并通向河畔露台和多条小径，白天看上去像是一条水晶的带子，夜晚则变成了一道光带。站在屋顶上、主楼梯或高处，还可以一览历史名城及其教堂的壮观景象。

舞台塔外面包有不锈钢网层，由距离塔顶 6 米的呈表面风化状的银色阳极氧化铝嵌板所支撑，这使得剧院外观看起来失去了物质的实体效果，其表面闪闪发光，随着白天天空及日落光线的变化而熠熠生辉。剧院的内部，挑高两层的门厅和主扶梯直通观众席。黑核桃木质地的观众席共有三层，并置有 1200 个座位，每个座位都包着夺目的红色／橙色皮革。Marlowe 演播室是个稍小的灵活的表演区域，可容纳 150 位观众。它建在门厅之上 6 米的位置上，视线通透，既可以一览河岸露台和河畔的风景，又能观赏到大教堂。像酒吧、咖啡厅这样的公共区域，可以通过全天候开放的河滨步道与走廊到达，在这里，游客一年四季都可以欣赏到节目和展出。

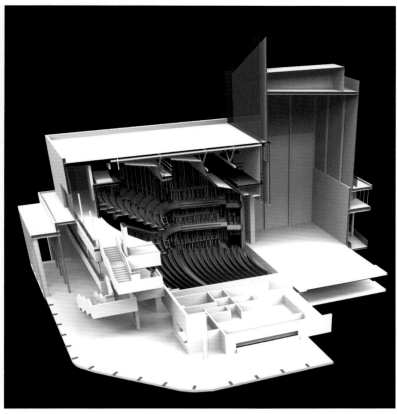

0 1 2 3 4 5   10m

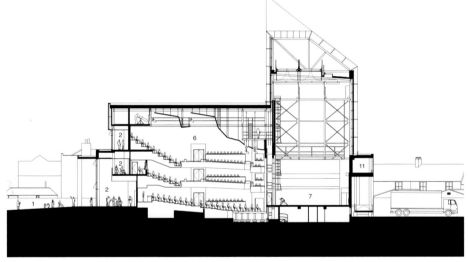

KEY

1. Paved Forecourt
2. Foyer
3. Box Office
4. Bar
5. Cafe
6. Auditorium
7. Stage
8. Dressing Rooms
9. Second Space
10. Creative Space
11. Administration Offices
12. Meeting Room

• Section AA

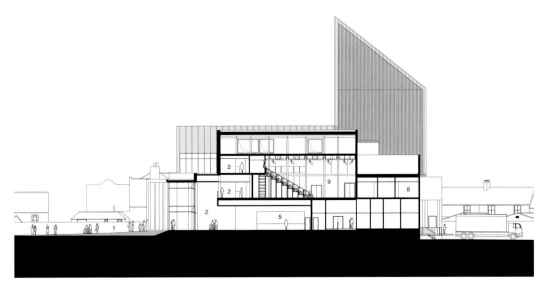

KEY

1. Paved Forecourt
2. Foyer
3. Box Office
4. Bar
5. Cafe
6. Auditorium
7. Stage
8. Dressing Rooms
9. Second Space
10. Creative Space
11. Administration Offices
12. Meeting Room

• Section CC

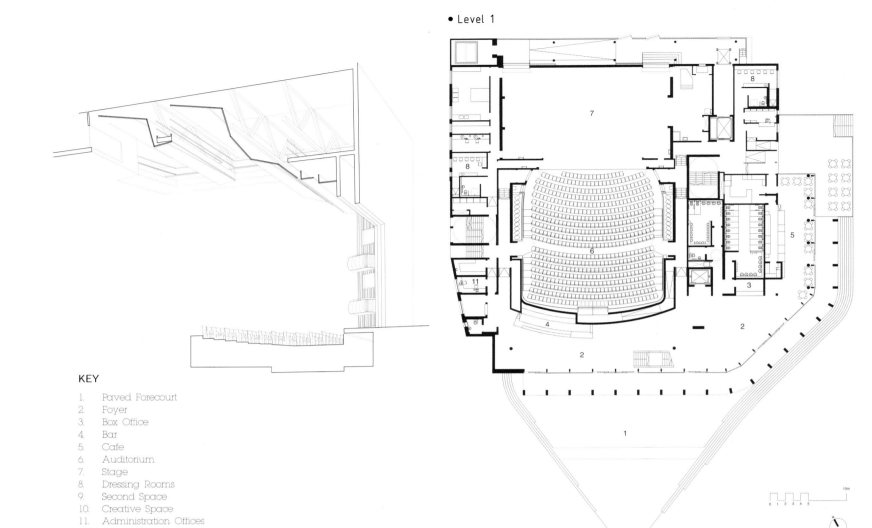

• Level 1

KEY
1. Paved Forecourt
2. Foyer
3. Box Office
4. Bar
5. Cafe
6. Auditorium
7. Stage
8. Dressing Rooms
9. Second Space
10. Creative Space
11. Administration Offices
12. Meeting Room

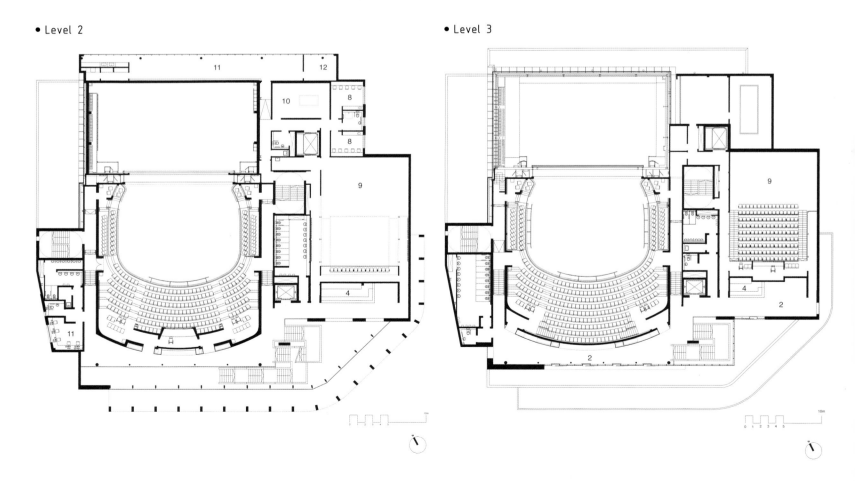

• Level 2

• Level 3

LOCATION_Miami, Florida, USA

SOUTH MIAMI-DADE CULTURAL CENTRE

Architect_ARQUITECTONICA
Associate Architect_The Hall Group
Interior Designer_ARQUITECTONICA
Theatre Planning and Design_Fisher Dachs Associates
Landscape_Curtis & Rogers Design Studio
Area_6,643m²
Photographer_Robin Hill

The facility consists of a 6,643 sq.m (71,504 sq.ft) cultural arts centre that includes a 3,159 sq.m (34,000 sq.ft) stage-house with 1,579 sq.m (17,000 sq.ft) front-of-house and public lobby space. The centre is a 966-seat facility that is intended to be used as a multi-purpose community centre to stage theatre and orchestral productions as well as local functions such as graduations and school plays. The back-of-house consists primarily of staff accommodations, building services, administrative offices as well as receiving and storage facilities. The activity building comprises an 700 sq.m (7,538 sq.ft) structure with 6.1m (20') high ceilings for the gallery, dance rehearsal and classroom spaces. It is intended that these can be used for local community meetings and after-hours adult education classes. The two buildings will be joined by an outdoor promenade leading to a gently sloping lawn for outdoor concerts and festivals along the Black Creek Canal. Outdoor activities along the canal edge will aid in the activation of the waterfront in tandem with the Park and Recreation. The design for this new cultural arts and community center is based on movement. The buildings reflect the spirit of movement, moving the patron through a visual as well as physical experience, making the patron a performer. The flow of people begins with the monumental ramps at the exterior of the building and continues through to the brushed aluminum grand-stair that delivers each person to the orchestra level and balcony levels above. Glimpses of people circulating behind a panelite-screened wall punctured with openings of various sizes at all the balcony levels, reinforces the idea that the building is designed with two prosceniums. The obvious proscenium is the one located within the performance hall as the traditional stage is set; the exterior frame around the full height curtain wall forms the second, subtler proscenium.

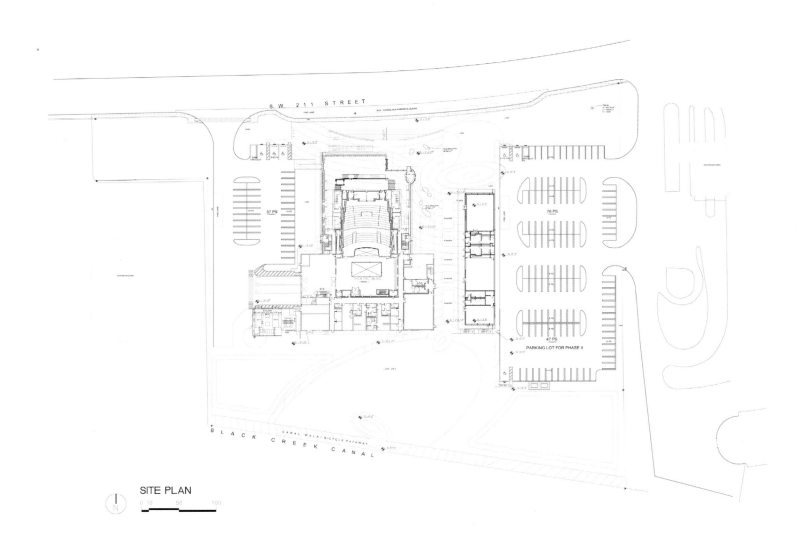

SITE PLAN

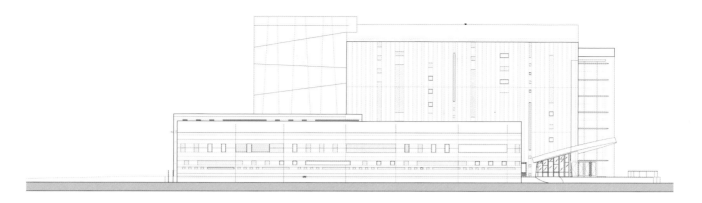

EAST ELEVATION

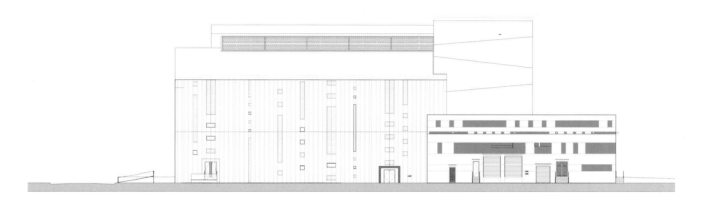

WEST ELEVATION

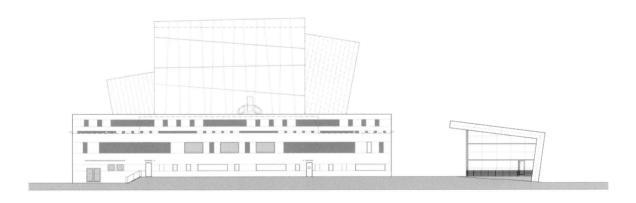

SOUTH ELEVATION

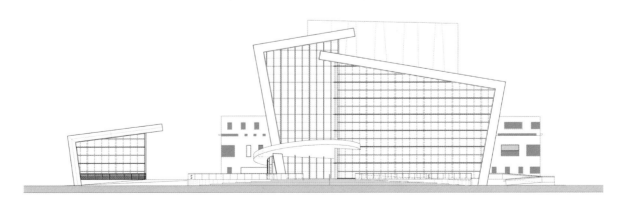

NORTH ELEVATION

171

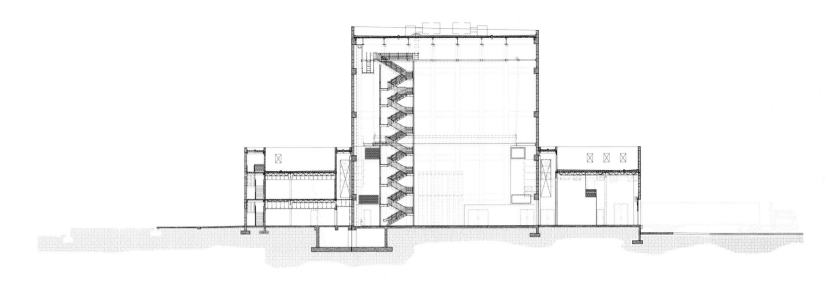

SECTION NORTH / SOUTH

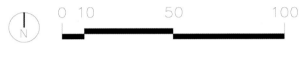

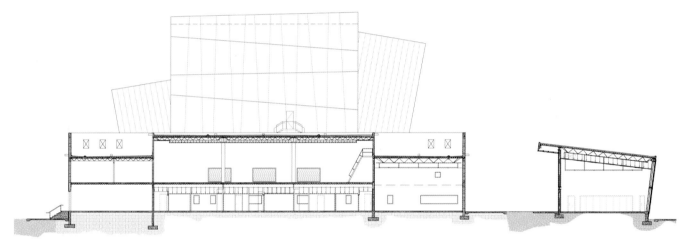

SECTION EAST / WEST B.O.H.

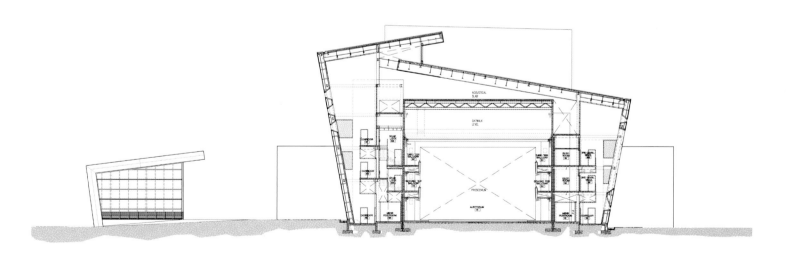

SECTION EAST / WEST

　　文化艺术中心占地6643平方米（71504平方英尺），包括公共休息厅和一个占地3159平方米（34000平方英尺）的舞台用房，其中含有面积为1579平方米（17000平方英尺）的前厅。建筑中心可容纳966个座位，计划作为多功能社区中心，用于举办戏剧、乐队演出，还可以发挥地方性功能，用来举办毕业典礼及学校活动。后台区主要包括员工食宿区、建筑服务区、管理办公室和储物室。活动厅是一个占地700平方米（7538平方英尺）、高6.1米（20英尺）的空间结构，用作画廊、舞蹈排练室和教室。这些都可以作为当地社区集会和成人教育的场所。一条户外散步街将两座建筑连接在一起，人们可以在黑溪运河沿岸舒缓的草坡上举行户外音乐会和节日庆祝活动。运河沿岸的户外运动将会带给海滨区和公园娱乐项目新的活力。这个新的社区文化中心的设计源于动感，建筑也折射出运动精神，给观众带去视觉与触觉上的双重体验的同时，让他们也成为表演者。人群走过建筑外围的坡道，登上镀铝的阶梯，到达上层的管弦乐演奏厅和楼厅。屏障墙后面来回走动的人影被墙上大小不同的镂空映射到楼厅的不同位置，更突出了剧院两个舞台的设计。表演厅内部最引人注目的是保留了传统阶梯模样的舞台，而外围构架则被落地窗帘环绕成了另一个淡雅的舞台。

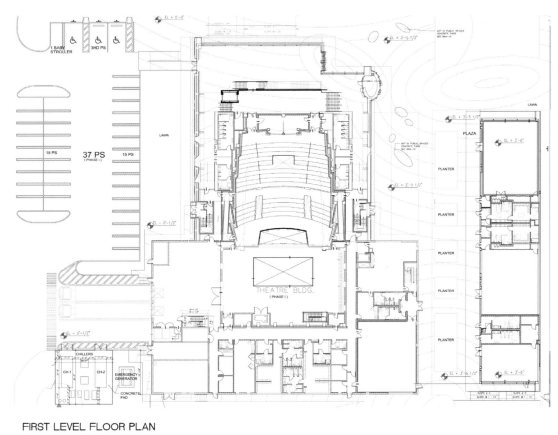

FIRST LEVEL FLOOR PLAN

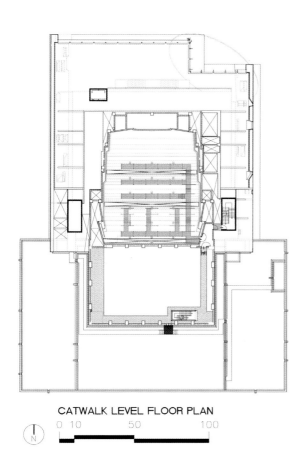

CATWALK LEVEL FLOOR PLAN

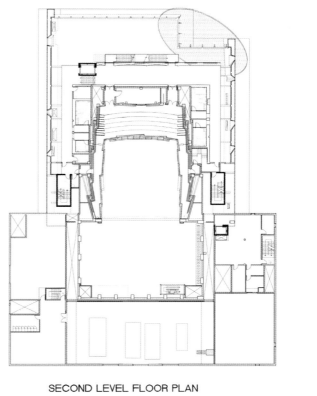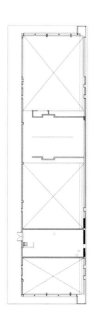

SECOND LEVEL FLOOR PLAN

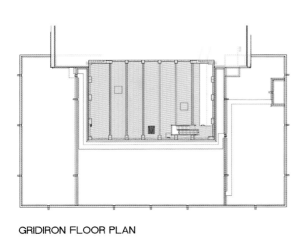

GRIDIRON FLOOR PLAN

LOADING GALLERY FLOOR PLAN

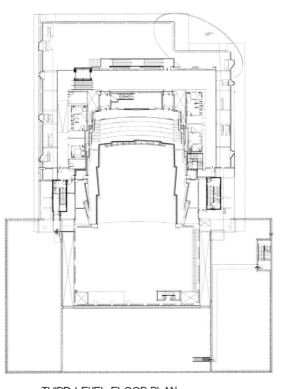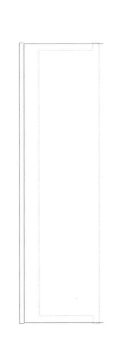

THIRD LEVEL FLOOR PLAN

LOCATION_ Montalto di Castro, Italy

NEW THEATRE IN MONTALTO DI CASTRO

Architect_Alessandro Corradini, Valerio Barberis, Marcello Marchesini, Cristiano Cosi
Firm_mdu architetti
Area_1,220m²
Photographer_Lorenzo Boddi, Valentina Muscedra, Pietro Savorelli

The design for the New Theatre in Montalto di Castro has a twofold objective: it is proposed as a conceptual model for measuring the territory and at the same time it attempts to express, through architecture, the magic of a theatrical event felt by the audience. The territory of Montalto di Castro sinks its origins into Etruscan enthronization whose ruins attest to architecture comprised of large stereometric masses in tufa; in the contemporary collective imagination Montalto di Castro evokes the world of the machines of the largest Italian power plant. The design proposes a temporal short circuit with respect to which the evolution of the territory is concentrated and expressed in a unique architectural moment: archaic Etruscan versus the aesthetics of the machine. The new theatre is a large concrete monolith characterized by subtle variations in color and texture, on which the fly tower appears to rest in an ethereal manner: an alveolar polycarbonate volume that dematerializes by day becoming indistinguishable from the sky, and lights up from within by night transforming into a large "lantern" on a territorial scale. A new, extended, piazza in travertine and concrete, designed as a diversion of the road providing access to the historic centre, leads to the entrance of the New Theatre identified by an impressive overhanging roof. It introduces visitors to a continuous environment in which the foyer and the auditorium flow freely into one another. The wooden walls, with their broken lines, create a space conceptually derived from the excavation of the concrete monolith. This morphological heaviness is contradicted by the vibration of the material that seems to envelope the space in a large curtain and introduces the spectator to the much awaited magical opening of the stage curtains. The auditorium that seats 400 has its counterpart in the outdoor arena that seats 500, which can thus benefit from the theatre stage.

蒙塔尔托-迪-卡斯特罗城的新剧院的设计有两个目的：一方面，它是作为测量地域的概念模型被提议创建的；另一方面，它要从建筑角度传递出观众对于戏剧魅力的理解。蒙塔尔托-迪-卡斯特罗区域最早可以追溯到伊特鲁里亚人统治时期，从其遗迹中可以发现，当时的建筑是用大块的石灰华立体结构构筑的。在积聚了所有对蒙塔尔托-迪-卡斯特罗的想象和猜测后，意大利的电厂里的机器唤起了世界的关注。在设计师创造的现世的短暂轮回中，蒙地的地域变迁通过独特的时代建筑被集中体现出来：古老的伊特拉斯坎人和机器美学的较量。新剧院是一个大型混凝土建筑，其颜色与质地的微妙变化分外抢眼，舞台塔好似屹立在半空中：一个聚碳酸酯成分的蜂窝状建筑在阳光的照耀下日渐与天空融为一体，而在夜晚，孔洞内的灯光直射向上则使其成为区域内的一个大"灯笼"。一个新的扩大的由石灰华和混凝土构造而成的走廊，是为了通向历史中心而设计的一条分岔路，直抵新剧院的入口。新剧院的屋顶则是一种给人以深刻印象的垂悬式屋顶。它给观众一种持续的空间感，在这里，大厅顺其自然地通向大礼堂。带有曲线美的木质墙给人一种延续于混凝土建筑的空间感。灵动的材料与形态上的沉重感形成鲜明对比，似乎是一个巨大的帷幕包围了整个空间，指引着观众来到舞台前，并期待神奇的舞台幕布的打开。礼堂能容纳400人，同时，室外表演场地还有500个座位，同样也可以欣赏到舞台剧。

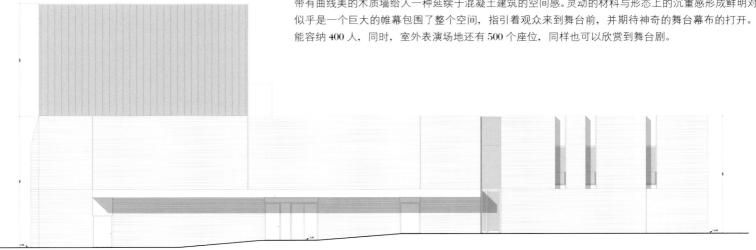

● East Elevation

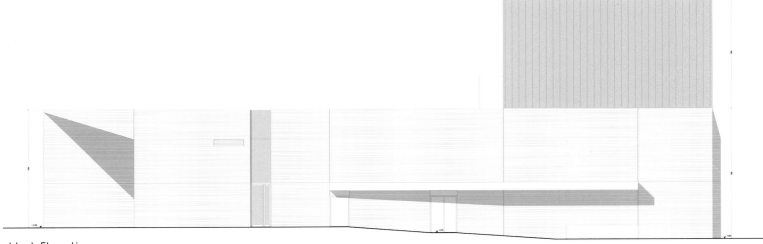

● West Elevation

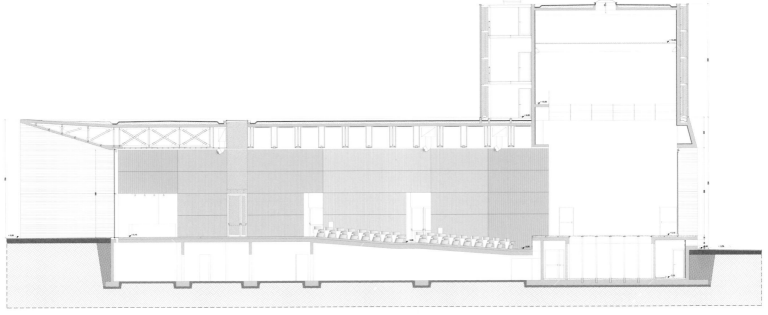

● Longitudinal section

185

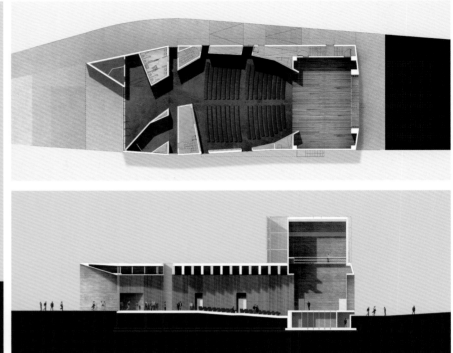
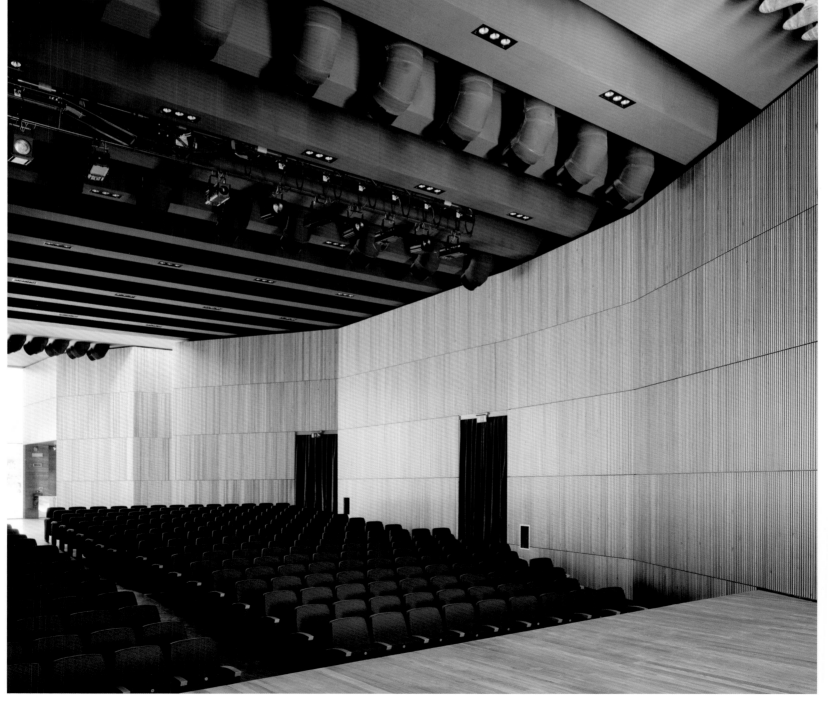

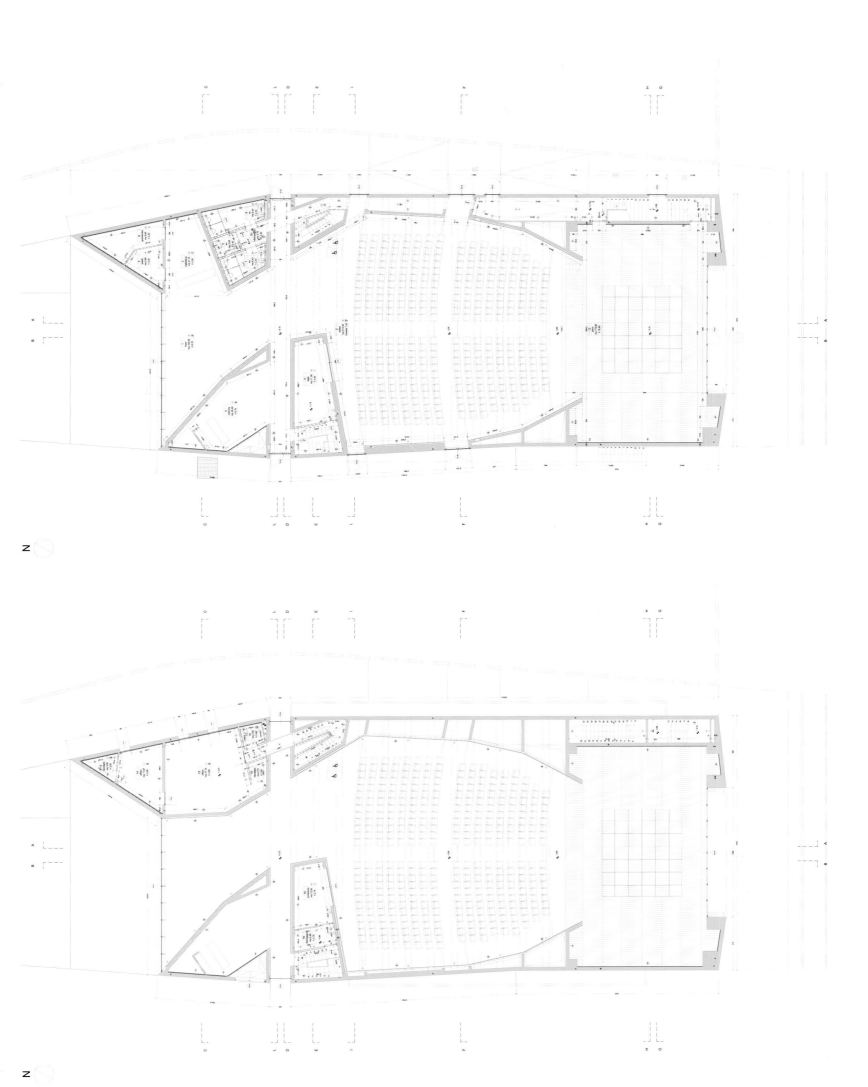

LOCATION_Burlington, Ontario

BURLINGTON PERFORMING ARTS CENTRE

Designer_Jack Diamond
Design Company_Diamond Schmitt Architects
Area_5,760m²
Photographer_Shai Gil
Director of Communications_Paul French

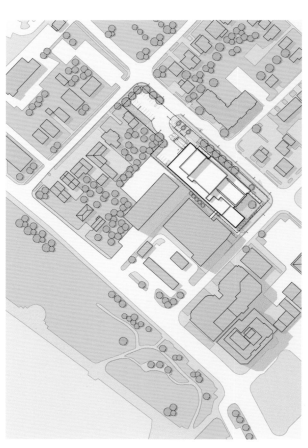

The city of Burlington, Ontario celebrated the opening of its new performing arts theatre today. A design team led by Jack Diamond, Principal with Diamond Schmitt Architects, created a 720-seat Main Theatre and a 260-seat Studio Theatre. A public plaza further integrates the facility with the community while the careful combination of materials, color, texture, lighting and architectural form brings unique architectural expression to downtown Burlington. "The entire building – inside and out – is activated as performance space to add a dynamic presence to the downtown neighborhood," said Gary McCluskie, Principal with Diamond Schmitt Architects. The wood-lined Main Theatre incorporates exemplary sightlines and excellent acoustics and provides the technical infrastructure for the most demanding performances. Other features of the Main Theatre include a six-storey fly tower, terrazzo flooring and a dramatically cantilevered balcony that seems to float in the room over the orchestra seating. The 5,760m^2 Burlington Performing Arts Centre is among the very first theatres in Canada designed with an aggressive sustainable directive to reduce energy consumption and lower its environmental impact. LEED strategies include sustainable site development, remediated contaminated soils, storm water management, reduced heat-island effect and reduced light pollution. Energy efficient design was achieved with thermal and lighting controls and monitoring, use of natural daylight, lighting fixture selection and exterior envelope design. A third party energy model indicates that the new facility will achieve a 77 percent reduction from the provincial energy use average and will meet the 2030 Challenge for environmental stewardship.

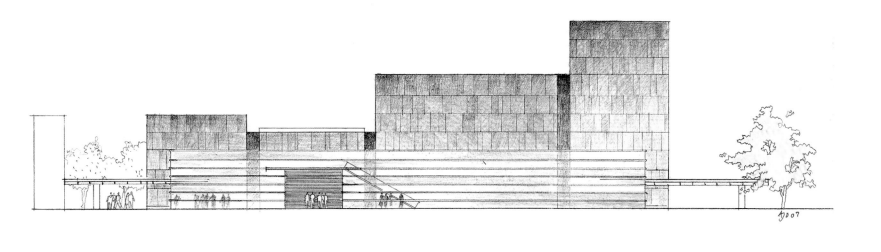

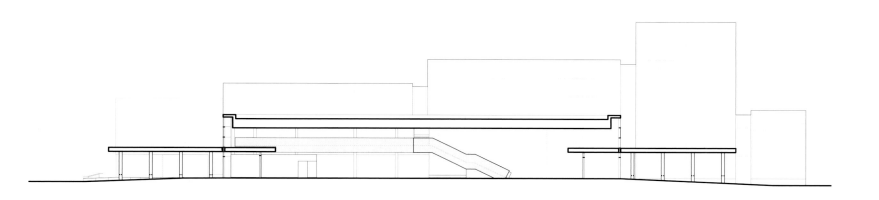

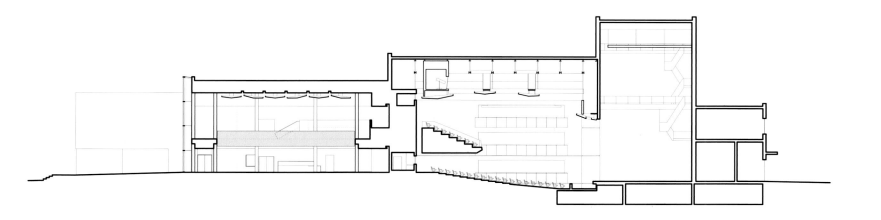

位于安大略湖区的伯灵顿市于今日庆祝其新型的表演艺术剧院的开业。由 Jack Diamond（Diamond Schmitt 设计师事务所负责人）所带领的设计团队，设计出了一个带有 720 个坐席的主剧场和一个内设 260 个座位的小剧场。一个公共广场进一步将艺术剧院与社区联系在一起。同时，材料、色彩、质地、灯光和建筑形式的巧妙结合也向伯灵顿市人民展现了一座独特的建筑。Diamond Schmitt 设计师事务所的另一位负责人 Gary McCluskie 说道，"整个剧院，无论室内还是室外，都作为表演空间被活跃起来，为市中心地带更添一份动感气息。"木质的主剧场拥有绝佳的视线和高品质的音质，为高要求的表演提供了技术设施。主剧场其他的特点包括：一个六层高的舞台塔、水磨石地面和夸张的悬臂式阳台，阳台看上去像是悬浮在乐队表演坐席的上空一般。占地 5760 平方米的伯灵顿表演艺术中心是加拿大最早的一批可持续性环保剧院之一，设计的最主要目的便在于减少能源消耗，降低对环境的影响。绿色能源与环境设计先锋策略包括地区可持续发展、土壤污染改善、积雨管理、减少热岛效应和减少光污染。节能设计的实现在于对光热的控制和监管、自然光的利用、照明设备的选择和外部空间的设计。第三方能源模型表明新设施的使用将会降低 77% 的城市能源平均消耗率，并能满足 2030 年环境管理提出的挑战。

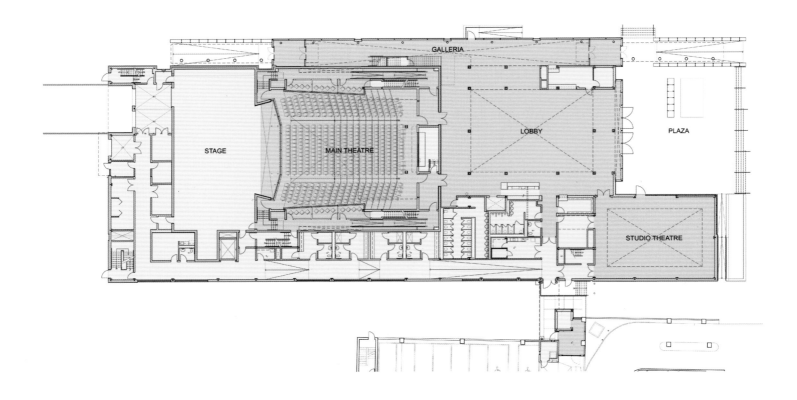
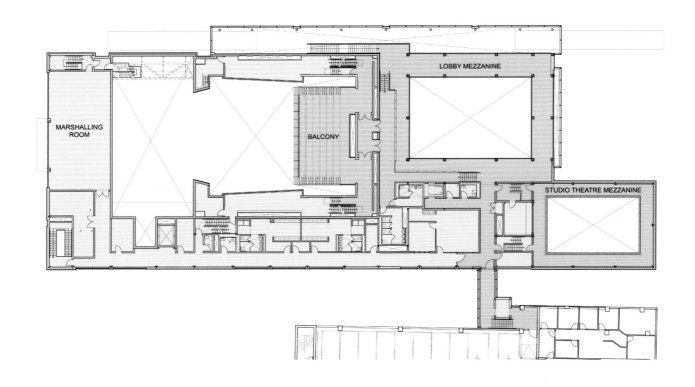

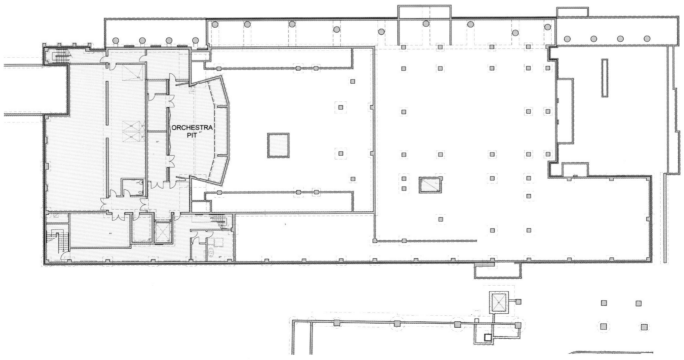

LOCATION_ Antonio Machado de Almonte, Huelva, Spain

TEATRO MUNICIPAL EN ALMONTE

Architect Author_ Juan Pedro Donaire Barbero
Architect Team_ Jesús Núñez Bootello, Carmela Domínguez Asencio, Tibisay Cañas Fuentes, David Rapado Moreno, Ignacio Núñez Bootello, Beatriz Hacar Hernández, Celine Nelke, Delia Pacheco Donaire, Pablo Baruc García Gómez
Photographer_ Fernando Alda, Kavi Sanchez
Website_ www.donairearquitectos.com

The building is located on the site of an old winery. It has the challenge of integrate the existing old buildings, declared as cultural interest, and being part of a cultural complex of a total of three buildings and a common space. This space turns into the main place of the town and an important meeting area.

An opportunity to work on light, material and space. The path chosen to work on these concepts, is the contrast. Contrast between outside and inside, between old and new, including a monumental scale and human scale. And the journey as the thread that sews and explains the intervention. A large area covered with large proportions and controlled height works as a high threshold. A monumental scale lobby welcomes the visitor showing the scale of a public building.

It is situated in the "Culture City" in Almonte, suh of Spain, next to the Rehabiltation of the Warehouse for use as a public library, and School of Arts (Music, Dance and Art) and enclosed within a public square. The intervention rigorously respected and maintained the different morphologies of each of the next buildings and this itself constitutes a strong characteristic of the project.

The materials are very important in the building, a opportunity to have a large experience working in innovated architecture with traditional and new materials. The concrete was used as the main material to contrast with the wood and this established the main line of work. The traditional Warehouse image contrast with the contemporary Theatre image, as a focal point for the public space. The brick contrast with the white concrete.

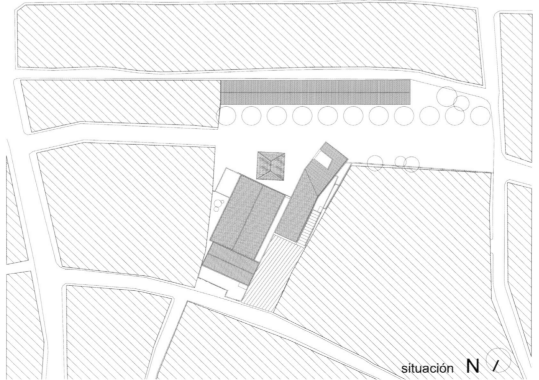

situación N

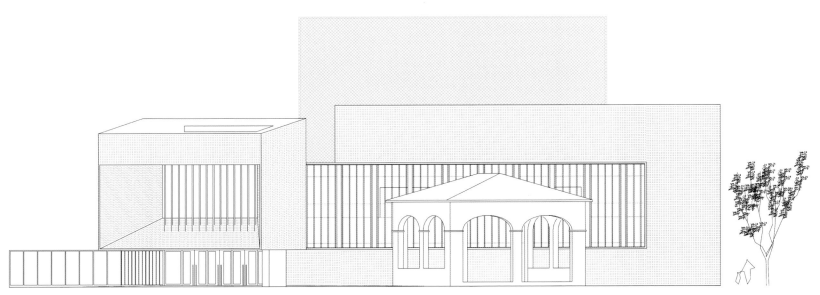

alzado principal
0 5 10 15m

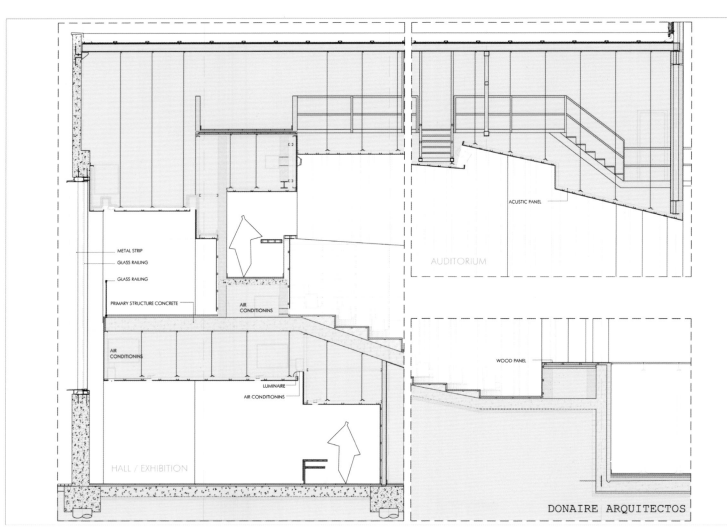

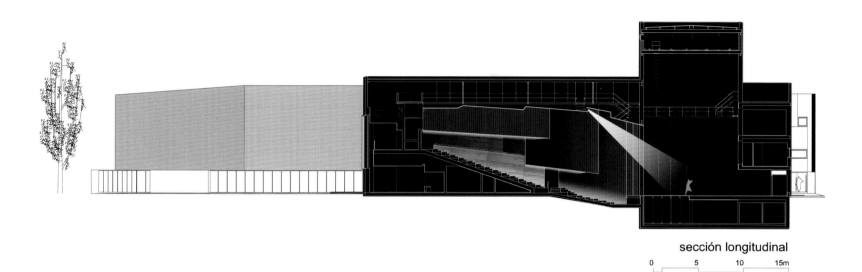

sección longitudinal

0　5　10　15m

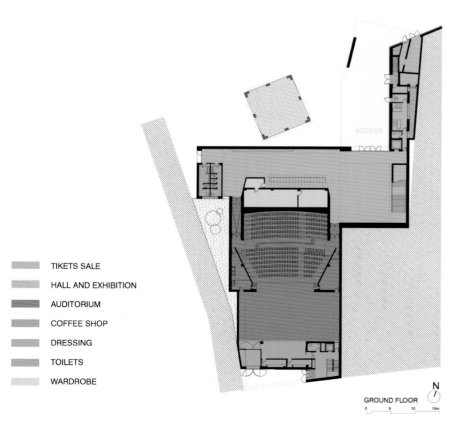

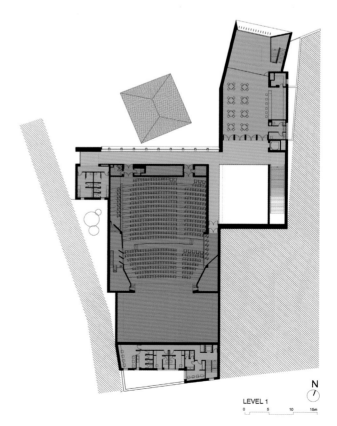

- TIKETS SALE
- HALL AND EXHIBITION
- AUDITORIUM
- COFFEE SHOP
- DRESSING
- TOILETS
- WARDROBE

GROUND FLOOR

LEVEL 1

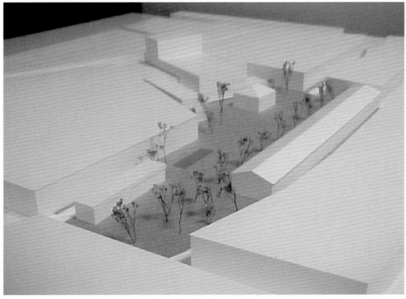
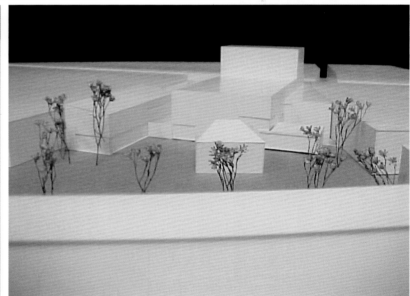

　　剧院坐落在一个老酿酒厂旁边。要整合被公认为是文物遗址的现存旧建筑，使之成为整个三个建筑体和公共空间文化中心的一部分是具有挑战性的。经过改造后，这个地方成为了城镇的会议中心和主要的活动场所。

　　这是一次集灯光、材料和空间为一体的工程。为了完成这一工程，设计上着重运用了对比。室内外的对比，新旧的对比，还包括纪念意义和人文内涵的对比。一连串的线索构成了引领人们进入室内的路线。入口处巨大的窗口式造型及其可控的高度显得蔚为壮观。有纪念意义的公共大厅迎接着来自各地的观众，向他们展示大规模的公共建筑。

　　剧院位于西班牙阿尔蒙特市的"文化城"内，毗邻复建后作为公共图书馆的仓库和艺术学校（教授音乐、舞蹈和艺术）。它同时还位于一个公共广场之内。这个工程极其尊重和保护其邻近的不同形态的建筑物，而这也是这个工程的特色之一。

　　建筑材料的使用是非常重要的，这个创新型建筑项目尝试运用了传统材料与新型材料。相比木材来说水泥是主要的建筑材料，它构成了整个建筑的主框架。与现代剧院形象相比，传统的仓库形象成为了公共区域的中心，砖块建筑与白色混凝土建筑形成了鲜明的对比。

LOCATION_Wenninkgaarde, Enschede, the Netherlands

OPERA HOUSE AND POP MUSIC STAGE ENSCHEDE

Architect_Jan Hoogstad
Architect firm_Ector Hoogstad Architecten
Area_18,000m²
Photographer_Jeroen Musch

This multifunctional music and theatre centre – comprising three halls, office spaces, studios and classrooms. Buildings featuring theatrical stages are among the most complex design projects. The main reason is that such venues are expected to be extremely flexible in order to accommodate a wide diversity of performances. The largest auditorium is technically equipped to accommodate plays and all sorts of dance and musical performances. The technical equipment and interior of Hall B ensure that, on a single night, this big pop venue can be magically transformed from a concert hall for all kinds of amplified music into a nightlife temple for DJ-hosted dance parties. In addition, Hall B can be used for small theatre performances and shows that require no stage. Backstage, the three halls share a spacious and efficient loading area, and the extra space thus created facilitates the assembly and dismantlement of stage sets, even in the case of complex productions. The concept shows three zones: backstage zone, play zone, audience zone. Each zone has its own atmosphere. It offers a comfortable working environment that performers find pleasant and are glad to come back to, combined with a well-oiled machine that offers a wide range of options. The building plays an important role in such experiences, and all kinds of resources – colour, material, texture, decoration – are employed to create an attractive experience. At night, a sophisticated lighting system is able to change the architecture of the three foyers, which form an entity during the day, into individual interiors with distinctive ambiences. Whereas the theatre foyer often provides a romantic and festive environment with an open character, its accompanying auditorium set among various levels like a jewel, the foyers of pop venues are usually more abstract and introvert, with lighting that visually lowers the space and directs attention to floor and skirting board.

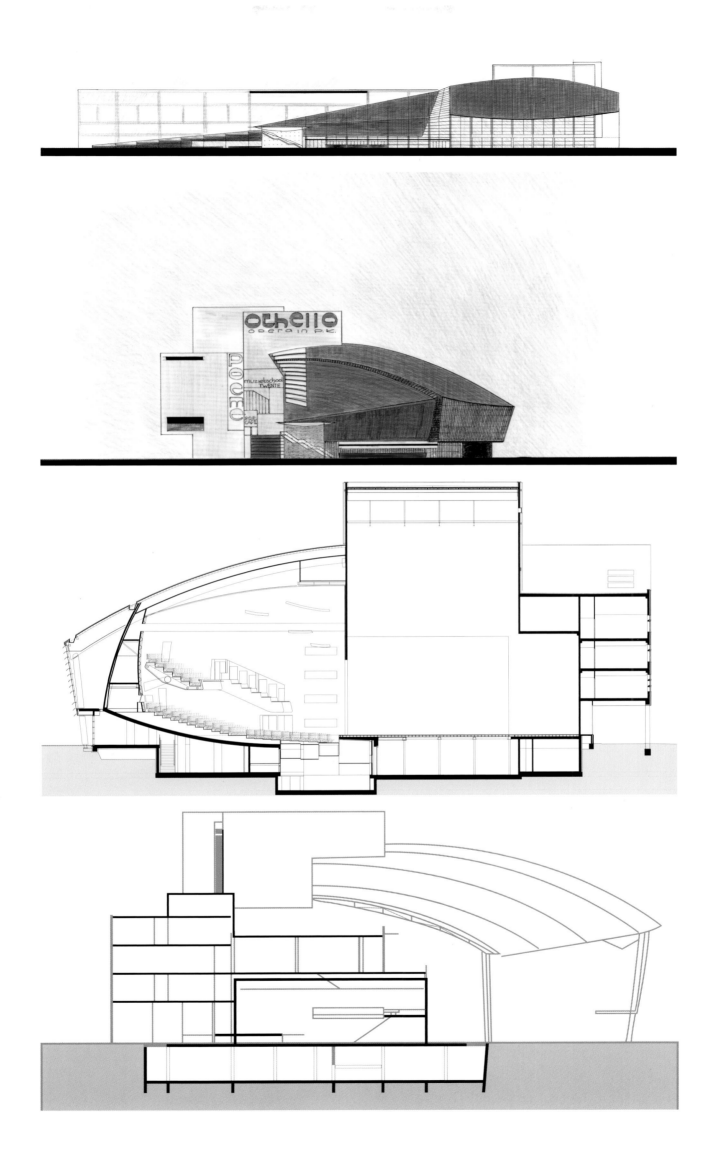

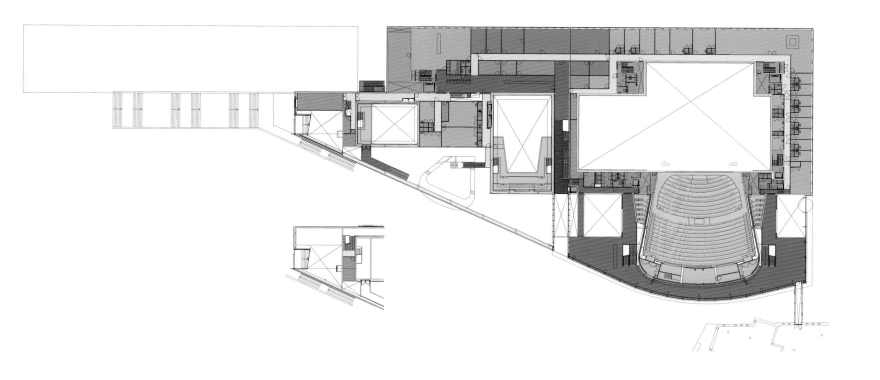
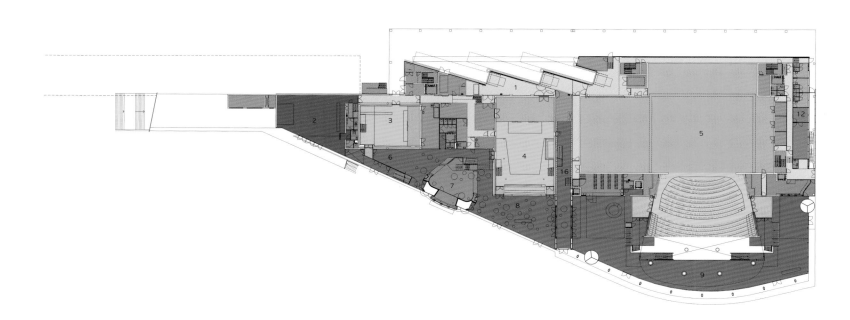
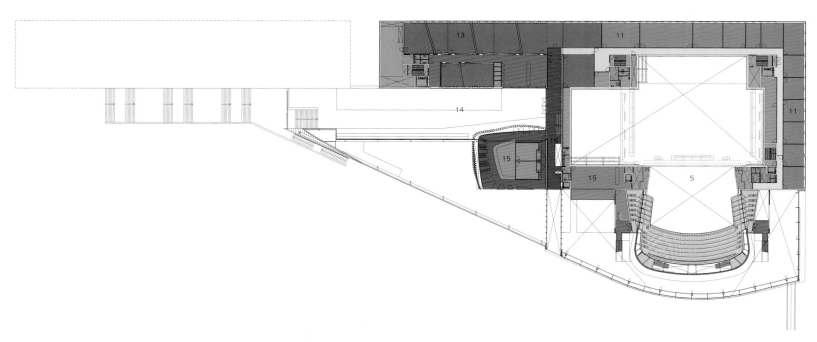

213

这个多功能音乐戏剧中心包括：三个礼堂、办公区域、工作室和练习室。具有戏剧舞台特色的建筑是最复杂的设计项目，因为此类场地设计要求具有极大的灵活性以适应不同种类的演出需求。最大的礼堂使用先进的技术设备以适应戏剧和所有种类的舞蹈和音乐表演。礼堂 B 的技术设备和内部设计确保了在夜晚表演时，这个大的流行场地可以奇迹般地从播放各种音乐的音乐厅转变成为夜场 DJ 主持的舞会。此外，礼堂 B 也可以用于小型剧场演出和不需要舞台的表演秀。在后台，三个大厅共用一个宽敞方便的等候区，这样剩余的空间就为场景和道具布置提供了集散的场所，即便是复杂的道具也可以。设计概念从三个区域：后台区、娱乐区和观众区体现出来。每个区域都有各自的氛围。它提供了舒适的工作环境，让表演者找到工作的乐趣并愿意回来工作，运转状况良好的设备也为其提供了多种多样的选择。这样，剧院的重要地位也被体现出来，各种资源如色彩、材料、质地和装饰的使用营造出引人入胜的感觉。到了晚上，复杂的照明系统可以把剧院分成三个大厅，这样白天的实体剧院就变成了几个具有独特氛围的个体区域。剧院大厅通常给人一种开放而浪漫的节日氛围，它旁边的礼堂像珠宝一样镶嵌在各层之间。音乐会场的大厅通常更为抽象、内敛，装饰的照明灯在视觉上降低了空间感，将人们的注意力引向地板和地脚线处。

LOCATION_Lleida, Spain

LA LLOTJA THEATRE

Architect_Mecanoo architecten, Delft
Local architect_Labb arquitectura S.L., Barcelona
Photographer_Christian Richters

The mountain with its historic cathedral Seu Vella and the Segre River mark the high point and low point of Lleida, the second city of Catalonia following Barcelona.

La Llotja theatre and conference center sits on the banks of the Segre, somewhat outside the centre of the city. Mecanoo's design interprets the landscape of Lleida as the exciting scenery before which the building has been placed, somewhat further from the river. The mise-en-scène is elaborated on three levels of scale. Regarded from the large scale of the region, the building forms a link between the river and the mountain. Viewed from the urban scale, La Llotja and the river form a balanced composition. At street level the cantilevers of La Llotja de Lleida provide protection from sun and rain.

Materials ensure distinction and orientation in the interior. The exterior is of stone. Inside there are mainly white, plastered walls and either wooden or marble floors. The entrance hall and the multi-functional hall have a marble floor, while the foyer has a floor of mixed hardwood. The theatre has the atmosphere of an orchard with walls of dark wood in which trees of light have been cut out. Thousands of leaves on the ceiling light the hall. The color palette of fruit is a theme that recurs in small details throughout the building. After all, the region of Lleida is famous for its fruit production. The roof is colorful: pergolas support a range of creepers and climbers like roses, jasmine and ivy. The garden with its mirador is not only pleasant but also useful since the roof cover keeps the building cool in the summer, provides a beautiful view for people living in the neighborhood and serves an extra place for conference guests to sojourn.

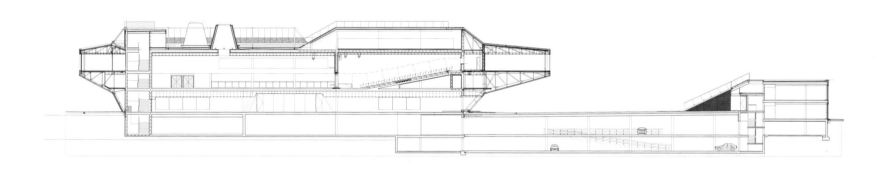

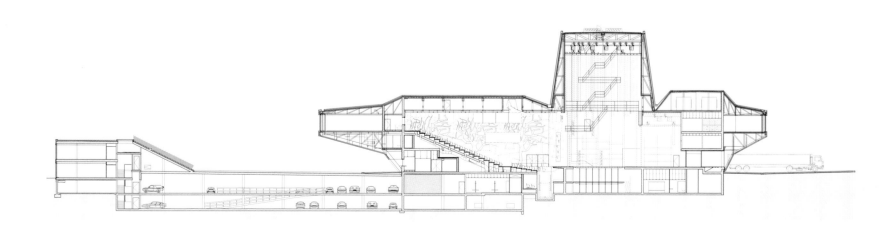

• Site Plan

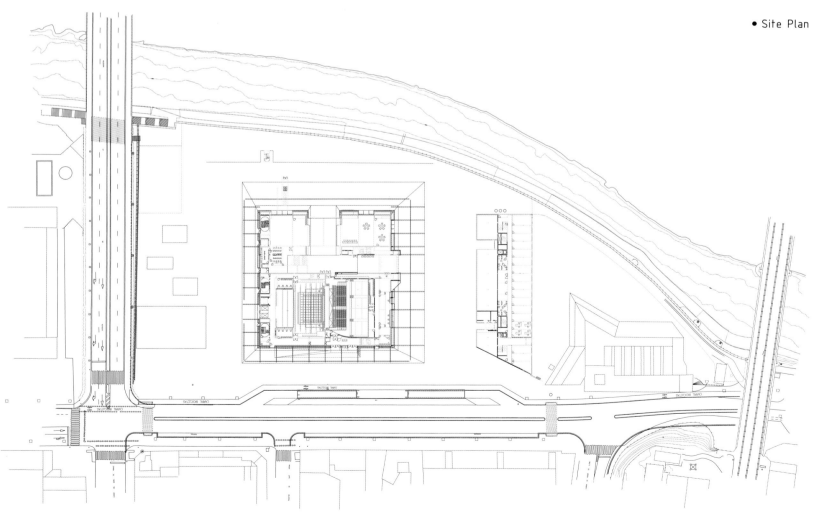

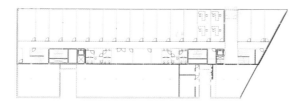
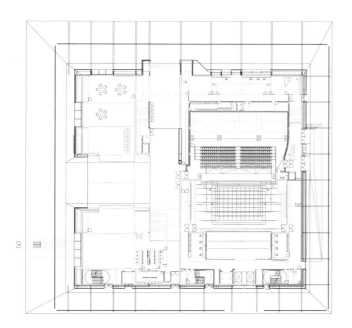

• Ground Floor

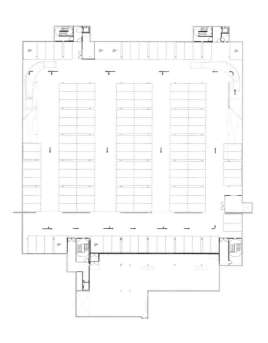

• -2 Floor

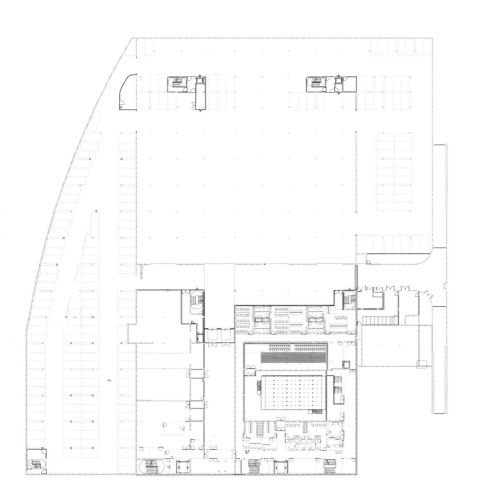

• -1 Floor

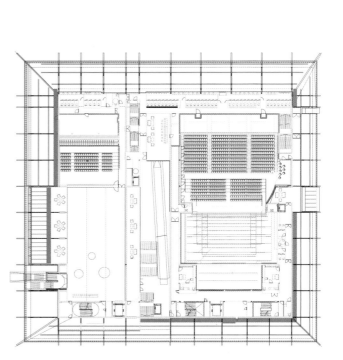

• 1 Floor

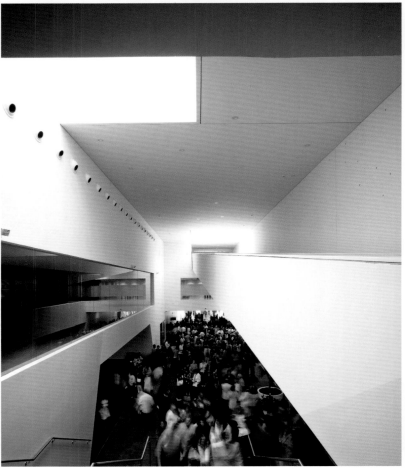

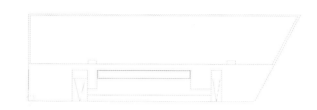

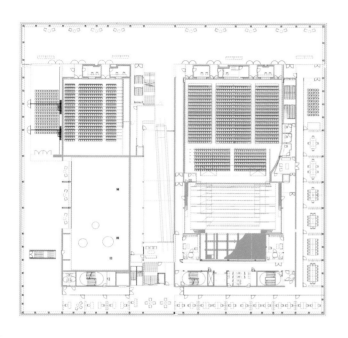

- 2 Floor

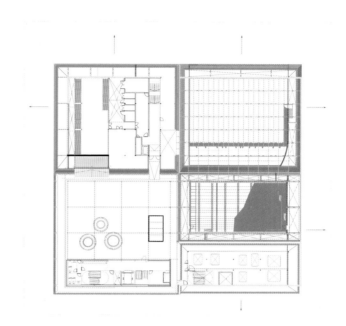

- 3 Floor

列伊达是仅次于巴塞罗那的西班牙加泰罗尼亚地区的第二大城市，其最高点是有着悠久历史的 Seu Vella 大教堂，最低点是 Segre 河。

城中心之外的 Segre 河岸边就是这个附带大型会议中心的 La Llotja 剧院的所在地。Mecanoo 设计师事务所认为列伊达地区的自然景观振奋人心，因此他们选择将此建筑建造在远离河流的地方。舞台演出在三个层面上得以诠释。从广义的范围上来看，这个建筑连接了河流与山脉；从城市的角度上来说，La Llotja 剧院与河流形成了一个均衡的复合体；从街区角度上来看，列伊达 La Llotja 影院的悬臂式设计则阻挡了强光和雨淋的破坏。

不同材料的运用确保了室内空间的独特性和区位关系。建筑外部材料使用石材，而建筑内部则主要是白色抹灰墙或者木 / 大理石地面。入口大厅和多功能大厅铺设大理石地面，而休息厅则是复合硬木地面。剧院有一种果园的氛围，深色的木墙上雕刻出树形的灯。天花上不计其数的叶子点亮了大厅。果实的色彩成为建筑中的一个主题，并在反复出现的微小细节中贯穿整个建筑。这样设计的原因也是因为列伊达因其生产水果而出名。建筑的屋顶是多彩的：棚架上爬满了玫瑰、茉莉花和常青藤等攀缘类植物。因为在夏天，屋顶覆盖物使得建筑里面很凉爽，所以带有眺望塔的花园不仅令人兴奋而且用处颇大。同时，它还给该区域的居民提供了一个美丽的景点，也使会议来宾多了一处逗留点。

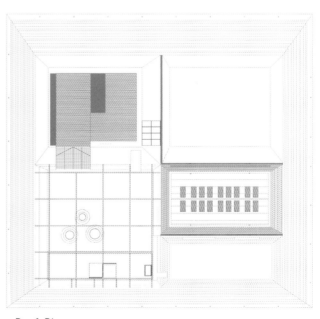

- Roof Plan

LOCATION_Century Avenue, Shanghai, China

SHANGHAI ORIENTAL ART CENTRE

Design Company_Paul Andreu Architecte, ADPi and ECADI
Area_39,694m²
Photographer_Paul Andreu agency

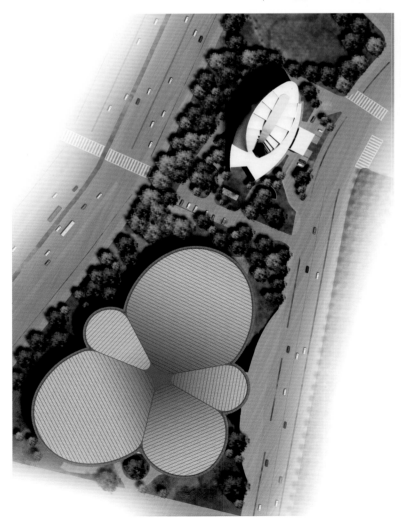

The Oriental Arts Centre project is a first rank public cultural building, encompassing mainly three venues: a 1,979 seats Philharmonic Orchestra Hall, a 1,054 seats lyric Theatre, a 330 seats chamber Music Hall. It also features ancillary public facilities, such a Exhibition Hall, Music Shops, restaurant and Arts Exchange premises : arts Library, Multimedia and training Centre. The project includes also all the suitable backstage facilities for the needs of the performance control areas, performances premises such as dressing rooms, rehearsal rooms and lounges. With a 39,694 sq.m construction site area, the project is built on 7 main levels. The halls will emerge from the base as trees from the earth; The building should be covered and enclosed by one unique cantilevered roof, linked by curved glass walls to the base; Spaces inside the building are distributed around and form a central circulation and meeting point. This should apply to the public as well as to the performers and the VIP's; The public space should be open and adaptable in order to increase the potential of use of the building; The performers should be provided with an efficient and agreeable working space; The three performance halls should be different in form and use different materials; The outside walls of the three halls will use enamel ceramic as their main common material; The material of the façades will be a glass incorporating a perforated metal sheet of variable density; The façades design itself will express innovation, modernity and enhance the public areas; Although secondary in their functional importance, the public spaces will define the character of the building and its filting with the spirit of the time.

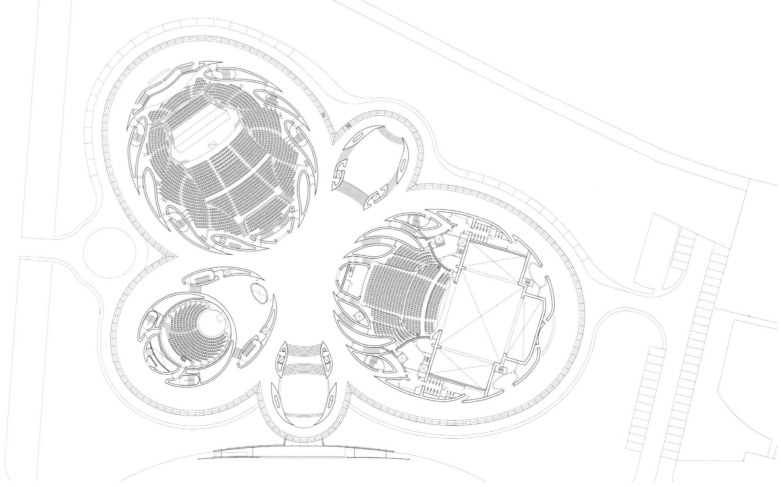

• Preliminary Design   Plan Level U2

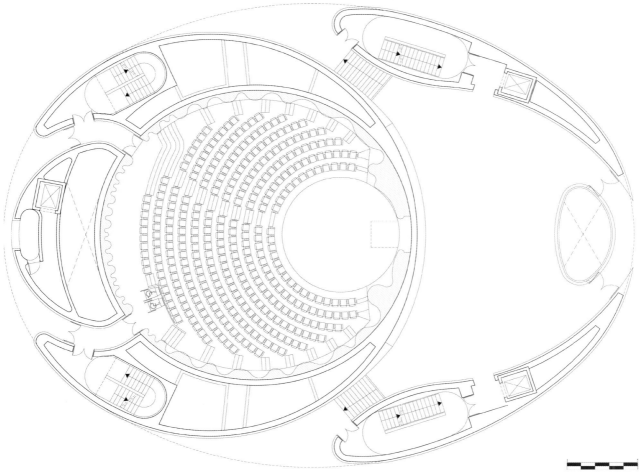

东方艺术中心是一座顶级的大众文化建筑,主要包括三个场馆:内设1979个座位的管弦乐厅,配有1054个座位的东方歌剧厅和设有330个座位的室内演奏厅。同时还具有配套设施,如:展览厅、唱片店、餐厅和艺术交流处(艺术图书馆、多媒体室和训练中心)。舞台控制区、化妆间、排练室及休息室等舞台场地所需要的各种后台设施一应俱全。该建筑占地39694平方米,注重七个方面的建设。大厅像树木一样拔地而起矗立在地表之上。独特的悬臂式屋顶环绕覆盖着建筑,由浮雕玻璃墙连着地面;内部空间环状分布,在中心形成了一个循环和交汇的公共场所,用来给大众还有表演者和贵宾享用;为了提高建筑的潜在利用价值,公共厅应既开放又具有调节性,为表演者提供更高效、宽敞的表演空间;三个演奏厅的构造形式和材料也不尽相同;三个演奏厅的外墙均将搪瓷釉作为主要材料;外立面则采用不同密度的金属夹层玻璃幕墙。外立面的设计传达着创新、现代的信息并凸显出公共区域;尽管在功能的重要性上居于第二位,公众区域却阐释了这座建筑的特性及它的时代精神。

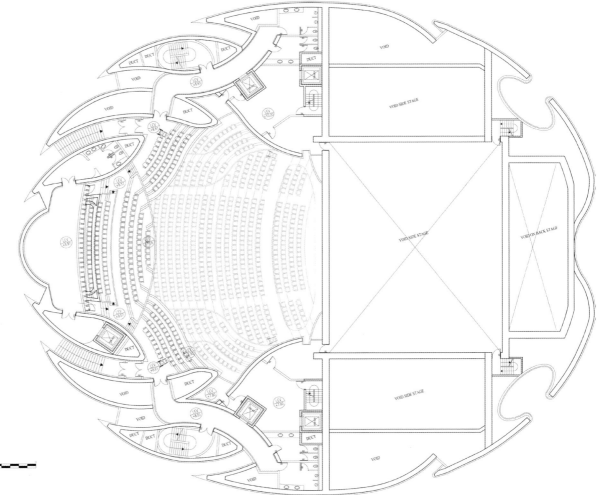

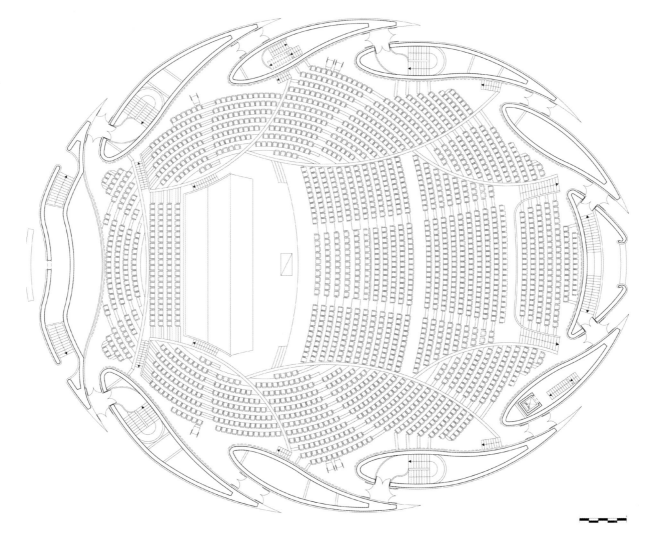

LOCATION_Angers, France

LE QUAI THEATRE IN ANGERS

Architect_AS.Architecture-Studio
Area_16,500m²
Photographer_LUC BOEGLY

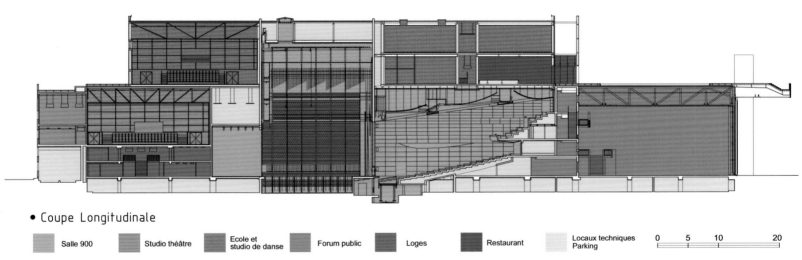

• Coupe Longitudinale

Salle 900 | Studio théâtre | Ecole et studio de danse | Forum public | Loges | Restaurant | Locaux techniques Parking

Le Quai rises on the banks of the River Maine, facing the Castle of Roi-René. It is the embodiment of the arts policy of the City of Angers Municipal Authority. Le Quai architectural competition, in its entirety from the conception of the plan to its realization, was awarded to Architecture Studio. Covering an area of 16,000m², Le Quai brings together five stage spaces: the theatre itself with a capacity of 900, and a smaller theatre with a capacity of 400. Two large rehearsal rooms, with a capacity of ninety-nine, will also be in operation. The foyer is the last interior space. It overlooks a wide grassy area between the covered road and the theatre building. The design of the building favours cultural diversity. It creates a dialogue between the City and the performing arts by providing open access to the foyer area and its exterior. The Complex is planned to embrace all types of audience and can readily adapt to the technical needs of contemporary performance. Thus, the project embodies the vision of the Council of the City of Angers and its Arts-for-All policy. In its concept of volumes and its choice of raw materials – glass, steel, concrete, AS.Architecture-Studio presents an ordered, albeit complex, interior, using a simple geometrical envelope. As soon as the spectator enters the building, the transparent façade of the foyer reveals a monochromatic interior but one which allows of additional spaces for varied and vivid colour compositions, in all materials.

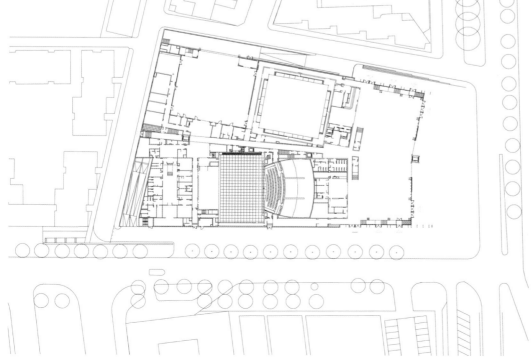

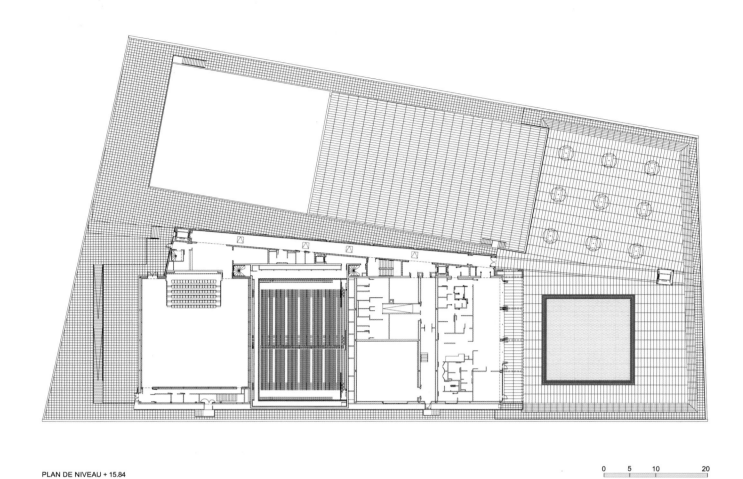

PLAN DE NIVEAU + 15.84

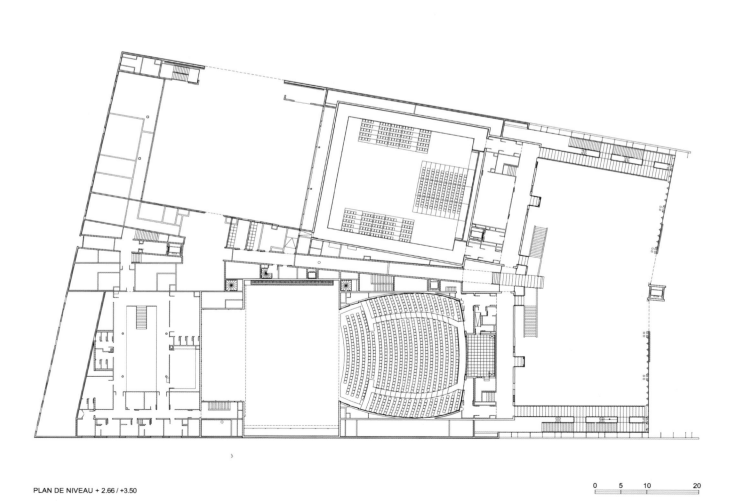

PLAN DE NIVEAU + 2.66 / +3.50

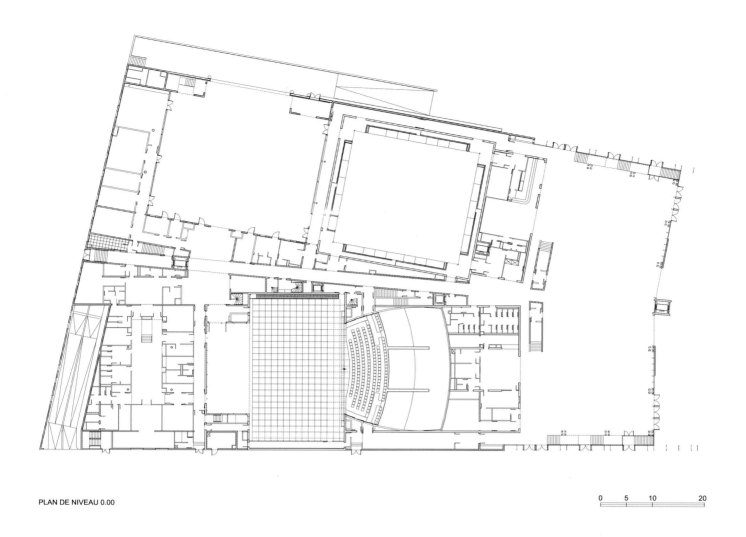

PLAN DE NIVEAU 0.00

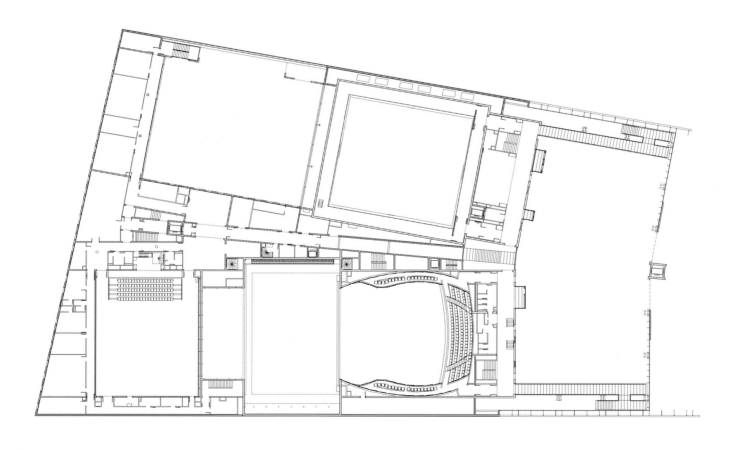

PLAN DE NIVEAU + 5.50 / +6.33

243

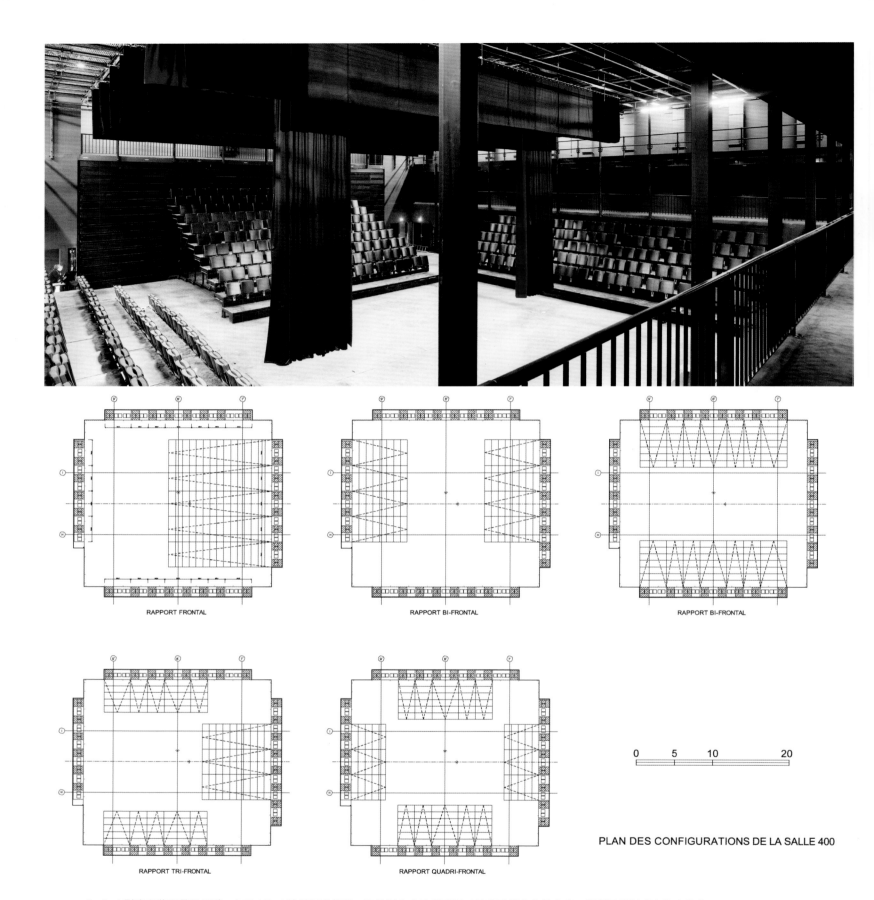

PLAN DES CONFIGURATIONS DE LA SALLE 400

Le Quai 剧院坐落于曼恩河畔,与 Roi-René 城堡隔岸相望。它是昂热市政管理局对艺术政策实施的化身。囊括了规划理念和实施的 Le Quai 建筑设计赛的大奖被 AS.Architecture 设计工作室获得。Le Quai 剧院占地 16000 平方米,共有五个舞台:剧院自身有 900 个座位,还有一个可容纳 400 人的小剧院。两个均可容纳 99 人的大彩排室也在建设中。前厅是最后的室内空间,它纵览了马路和剧院大楼之间的宽阔的草地。剧院的设计有利于文化的多样性,通过为剧院前厅和外界搭建平台,沟通了城市和行为艺术。这个复合建筑旨在满足所有观众的需求,并能快速适应当代表演的技术需求。因此,该设计项目体现了昂热市政府的远景和其艺术服务于一切的政策。在建筑理念和对原材料如玻璃、钢材、水泥的选择上,AS.Architecture 设计工作室体现出一种顺序感,内部选用一种看似复杂、实则简单的几何图形。一旦观众走进剧院,便能通过剧院透明的外观看到单色的内部结构,此种结构也凸显了由其他材料制成的多彩且多变的空间。

LOCATION_ Deventer, the Netherlands

DEVENTER SCHOUWBURG

Design Company_ M+R interior architecture

The Deventer Schouwburg is the city's stage, the meeting place for producers and visitors, and as much it's important that everyone feels welcome. So M+R designed an open and multifunctional interior. The transparent façade makes the large lobby on the ground floor seem like a continuation of the public area. With one theatrical movement the two high revolving doors, which are made of glass, sweep visitors onto the stage.

The bar, cloakroom and restaurant are positioned in such a way that the lobby can also be used for presentations or receptions and the like. Also along the stairs there are generous, organically shaped stages/sitting areas that form an integral part of this area. Large sliding doors allow the lobby to be used separately, enabling a more efficient use of the theatre. The style of the lobbies on the first and second floor has been modified and they have been given a new bar and a raised platform.

Deventer Schouwburg 剧院是整个城市的舞台，它不但为制片人和游客提供了聚会场所，更为重要的是，它能让每个人都感觉到自己是受欢迎的。所以 M+R 设计师事务所设计了开放、多功能的内部空间。剧院透明的外观让一层的大厅看上去像是公共区域延伸出来的一部分。两个高大的玻璃旋转门用一个夸张的动作便把游客"扫到"了舞台上。

酒吧、衣帽间和餐厅的设计方式也可以把大厅变成报告厅或是接待处之类的地方。沿着楼梯，有一个状似舞台/座位席的宽敞的区域，这是整个区域的主体部分。大扇的滑动门把大厅分成可独立使用的区域，这样能更有效地利用剧院。一层和二层的大厅风格经过修改后，新增了酒吧和升降台。

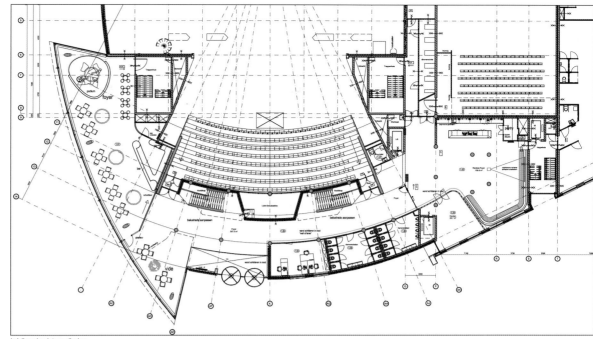
indelingsplan 1ste verdieping

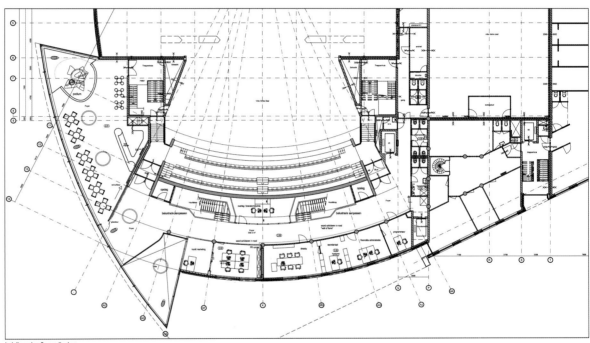
indelingsplan 2e verdieping

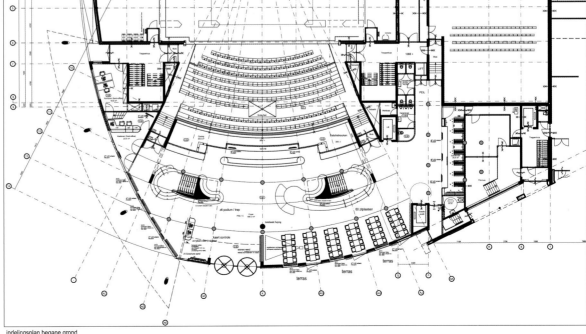
indelingsplan begane grond

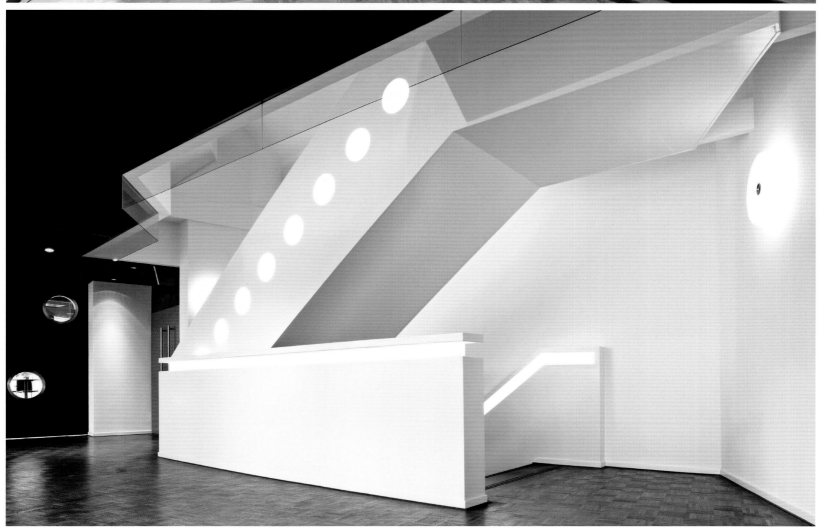

253

LOCATION_Athens, Greece

ONASIS CULTURAL CENTRE

Architect_Architecture Studio
Co-architect_AETER
Area_18,000m²

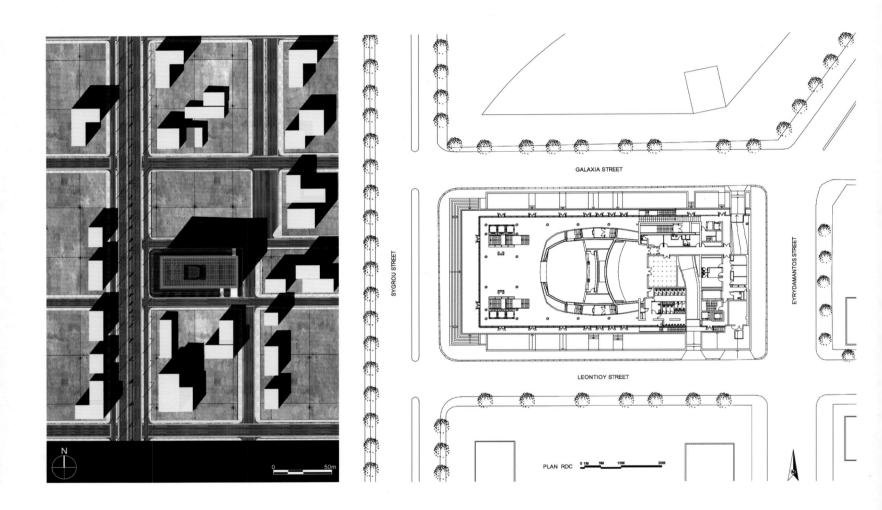

Foundation including an opera house – theatre of 900 seats, an auditorium conference hall – cinema of 200 seats, an open air amphitheatre of 200 seats, a library, a restaurant and an exhibition hall.

The building is designed as a simple, diaphanous volume built of Thassos marble, elevated over a glass base. The facades are simultaneously opaque and transparent depending on whether one perceives them from near or afar. The project is revealed in the movement, the approach towards the building. The opacity of the stone is balanced by transparency, rhythm and matter. This urban scenography is based on the treatment of the building's facades as a living, responsive membrane, reflecting the activity of the Foundation itself and its openness onto the city and the world. The simplicity of the volume and the abstraction of architectural expression create the monumental aspect that would have seemed almost impossible because of the small size of the plot.

Beyond the facades, a precious and surprising object that brings together the three auditoriums crosses the entire building. This is the heart of the project. Magnified by the void surrounding it, its envelope forms the ideal stage for the Cultural Centre's events, whose presence is marked at the scale of the building and that of the city.

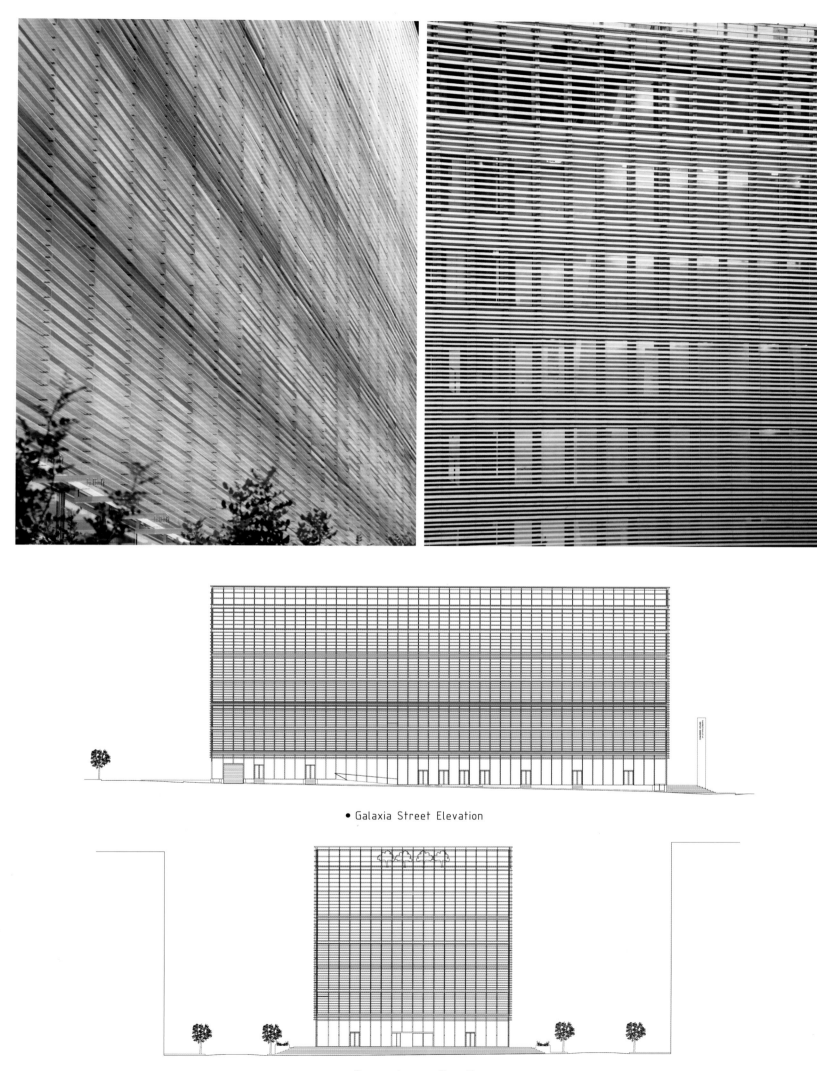

• Galaxia Street Elevation

• Sygrou Avenue Elevation

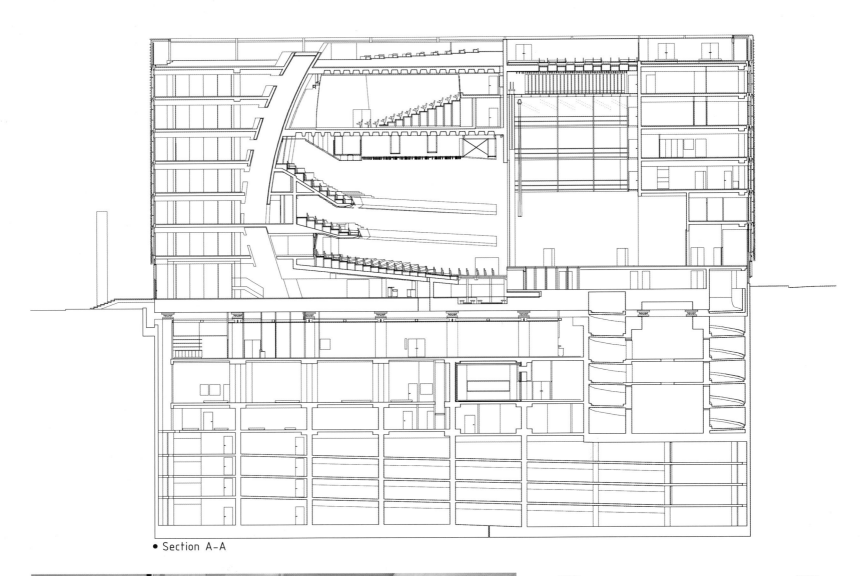

• Section A-A

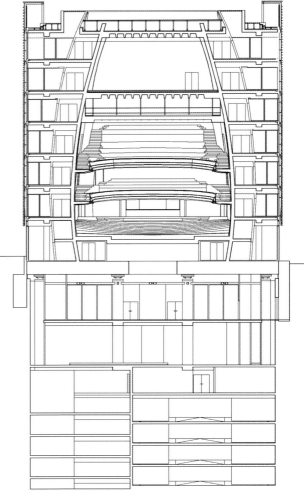

• Section D-D

剧院的设计包括一个可容纳900人的歌剧院,一个可容纳200人的会议厅、电影厅,一个可容纳200人的开放式圆形剧场,一个图书馆,一个餐厅和一个展览厅。

剧院设计精简,白水晶大理石和玻璃底板的使用使整个建筑精致透明,高大挺拔。建筑的外观看上去是否透明取决于人们看它时距离的远近。该建筑的设计内涵体现在动态中。不透明的石头与透明度、节律和材料相平衡。

城市的配景方法是把建筑外观作为生动的、能回应的一面镜子来反映建筑物本身的活动以及此建筑对城市和世界的开放度。建筑外观的简洁性及抽象性的建筑表情所营造出的永久感,并没有因为建筑的规模小而打折扣。

除了外墙,一个罕有的、令人惊讶的物体贯穿整个主体建筑,并把三个礼堂连接在一起。这是整个设计的核心部分,周边的空旷场地放大了这个核心部分,它的包围圈为文化中心的活动提供了一个完美的舞台,它的存在也标志着建筑的规模和整个城市的规模。

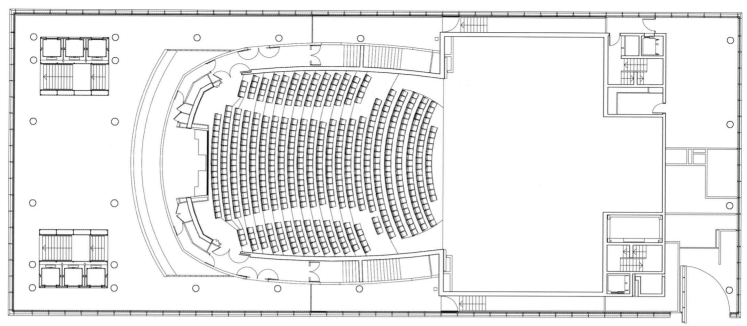

PLAN +1   Theatre 1

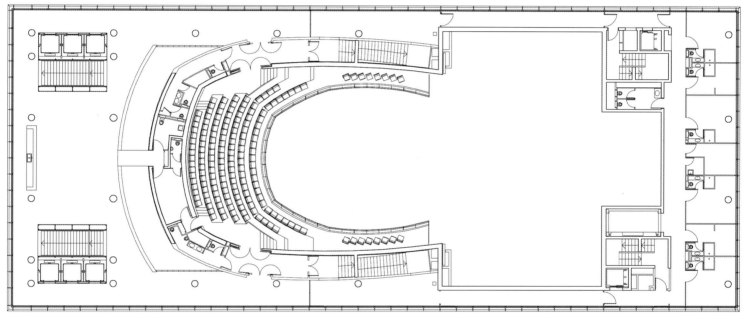

• PLAN +2   Balcon 1

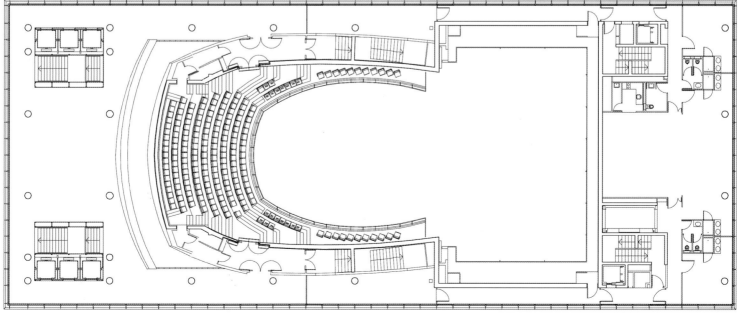

• PLAN +3   Balcon 2

LOCATION_ Bispetorv, Aarhus, Denmark

AARHUS THEATRE

Architect_ Hack Kampmann (1900, original building)

Firm_ C. F. Møller Architects (1955-2011 extensions, restoration, maintenance and refurbishment)

Photographer_ Julian Weyer

Aarhus Theatre was built in 1900 by the architect Hack Kampmann, and the original building only included the Theatre auditorium with the stage and few other facilities. Since 1955 C.F. Møller Architects has been involved in numerous extensions, restoration, maintenance and refurbishment work of the now listed building. The theatre distinguishes itself by its very characteristic "Art Nouveau" architectural style, with rich sculptural ornamentation by Danish artist Karl Hansen-Reistrup.

Most recently extensive conservation works was done in the main theatre space and modernized technical facilities were added. Also a new café and foyer stage was designed in the north wing of the building. At present an extensive renovation of the second largest stage at Aarhus Theatre, Scala, is taking place. At the same time, the public areas will also be brought up to modern standards as will the smaller stages: Stiklingen and the Studio-stage. Scala is originally an extension of the Aarhus Theatre designed by C. F. Møller Architects and built in 1955. Until 1980, Scala served as concert hall and movie theatre with 880 spectator seats. Since then, Scala has served as Aarhus Theatre's second largest stage, with almost 300 seats – like the rest of the theatre the interior is listed, including the lighting fixtures especially designed in 1955 by world-famous designer Poul Henningsen.

270

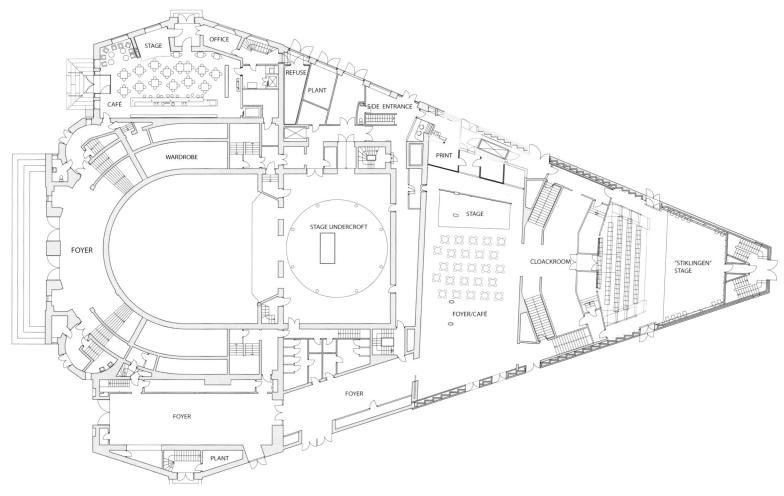

• Ground floor

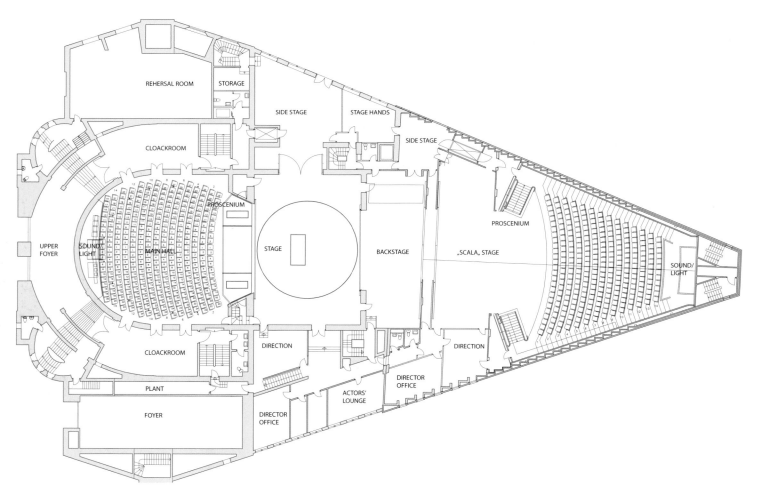

• 1st floor

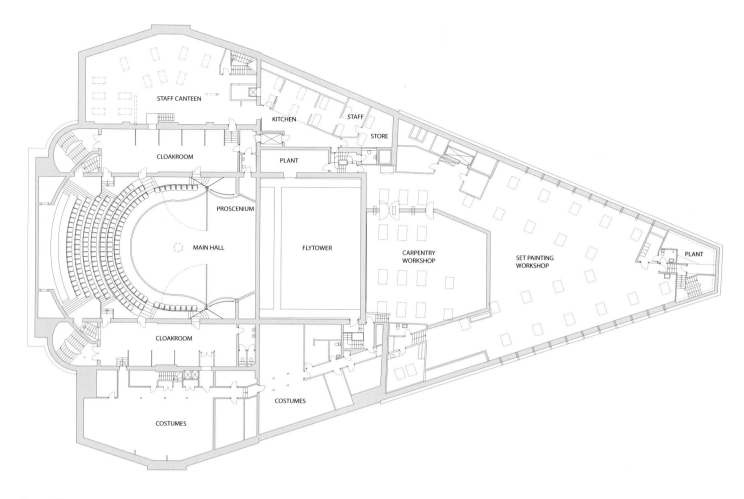

• Upper floor

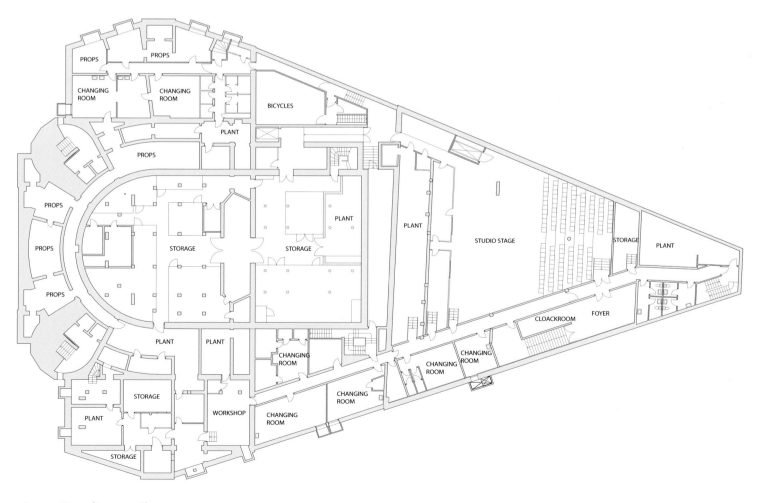

• Lower floor (Basement)

奥尔胡斯剧院由建筑师 Hack Kampmann 始建于 1990 年，剧院原始建筑只有观众席、舞台和一些其他设施。从 1995 年起，C.F. Møller 建筑师事务所便开始对这个现已是文物的建筑进行了多次扩建、翻修、维护和整修工作。奥尔胡斯剧院因其"新艺术派"的建筑风格而与众不同，大量的雕刻装饰物均出自丹麦艺术家 Karl Hansen-Reistrup 之手。

最近对剧院的主体空间做了大面积的整修工作，并添加了很多现代化的技术设施。同时，在剧院的北翼还设计了新的咖啡厅和门厅台阶。目前，奥尔胡斯剧院第二大舞台 Scala 的整体翻修工作也正在进行中。与此同时，公共区域以及 Stiklingen 和 Studio-stage 这两个小舞台也会以现代标准重建。Scala 舞台是 1955 年由 C.F. Møller 建筑师事务所为奥尔胡斯剧院扩建的一部分，直到 1980 年，Scala 才作为独立的音乐厅和电影院而存在，可容纳 880 位观众同时观看。自从那时起，拥有 300 个座位的 Scala 成为奥尔胡斯剧院的第二大舞台——像剧院的其他地方一样，Scala 也是被登记在册受到保护的，尤其是其内由著名设计师 Poul Henningsen 在 1955 年为其设计的照明灯具。

LOCATION_Lelystad, the Netherlands

THEATRE AGORA

Architect_Ben van Berkel
Firm_UNStudio
Area_7,000m²
Photographer_Iwan Baan, Christian Richters

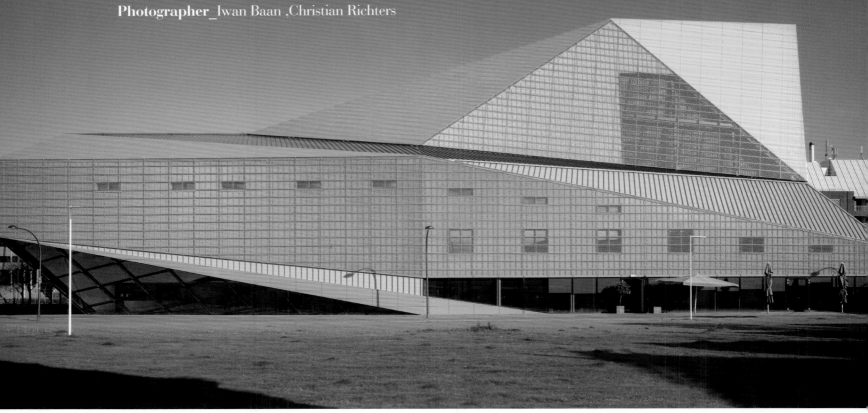

The Agora Theatre is an extremely colourful, determinedly upbeat place. The building is part of the masterplan for Lelystad by Adriaan Geuze, which aims to revitalize the pragmatic, sober town centre. The theatre responds to the ongoing mission of reviving and recovering the post-war Dutch new towns by focusing on the archetypal function of a theatre: that of creating a world of artifice and enchantment. Both inside and outside walls are faceted to reconstruct the kaleidoscopic experience of the world of the stage, where you can never be sure of what is real and what is not. In the Agora theatre drama and performance are not restricted to the stage and to the evening, but are extended to the urban experience and to daytime.

Inside, the colourfulness of the outside increases in intensity; a handrail executed as a snaking pink ribbon cascades down the main staircase, winds itself all around the void at the centre of the large, open foyer space on the first floor and then extends up the wall towards the roof, optically changing color all the while from violet, crimson and cherry to almost white.

The main theatre is all in red. Unusually for a town of this size, the stage is very big, enabling the staging of large, international productions. The intimate dimensions of the auditorium itself are emphasized by the horse-shoe shaped balcony and by the vibrant forms and shades of the acoustic paneling.

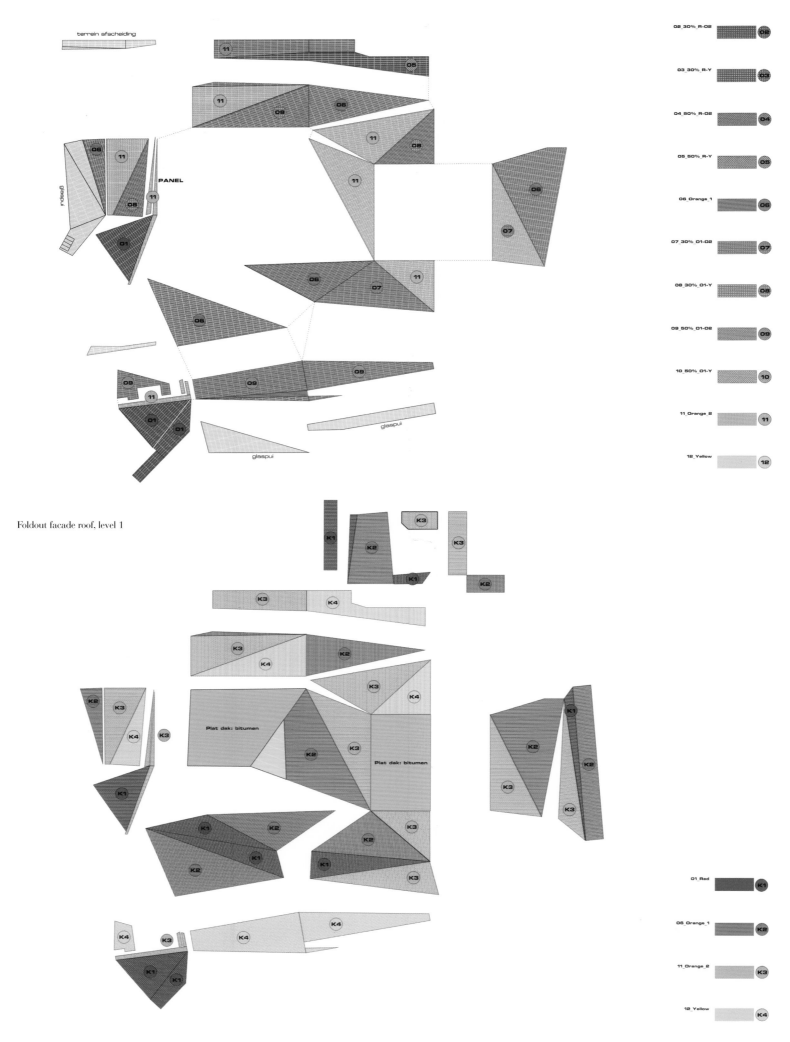

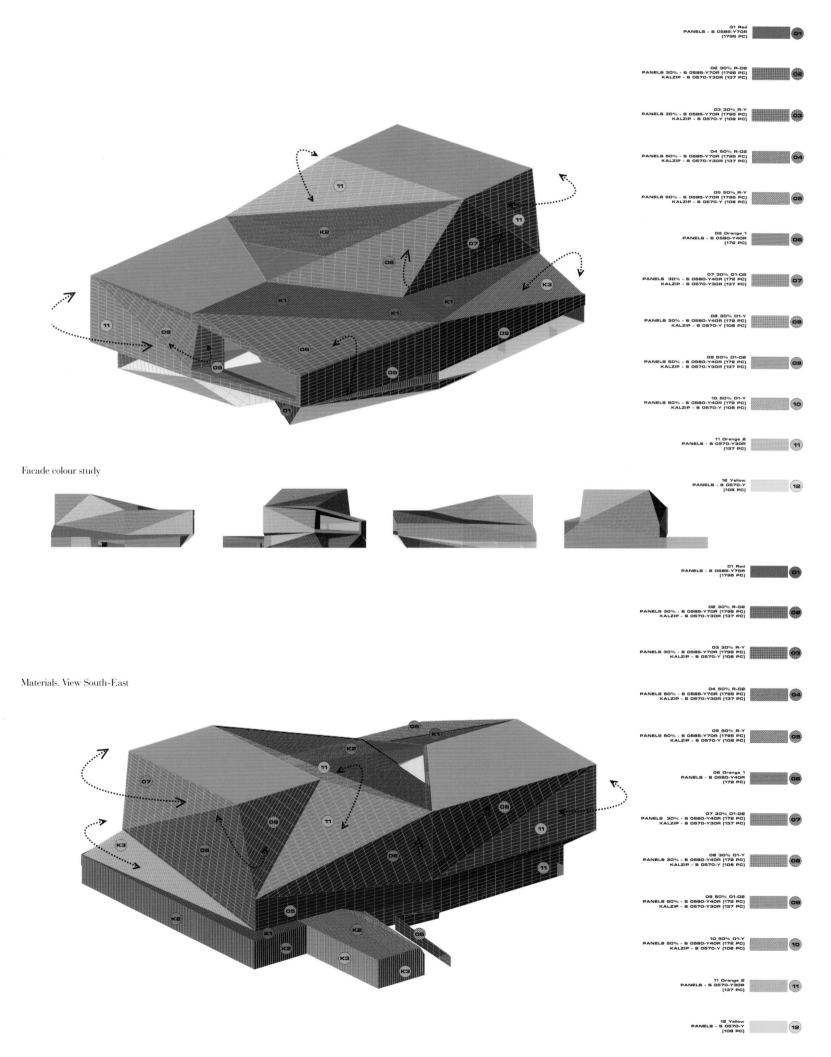

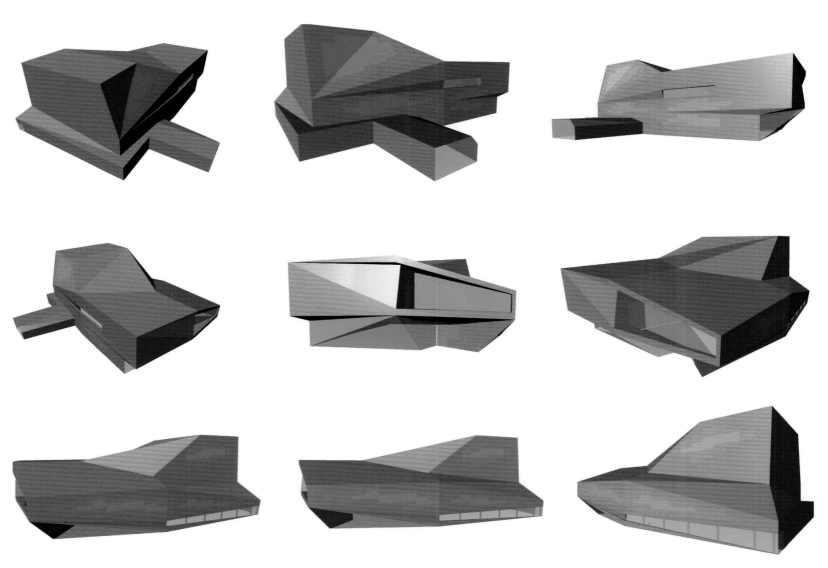

Diamond structure

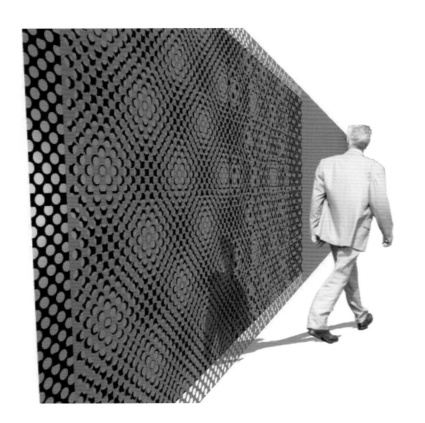
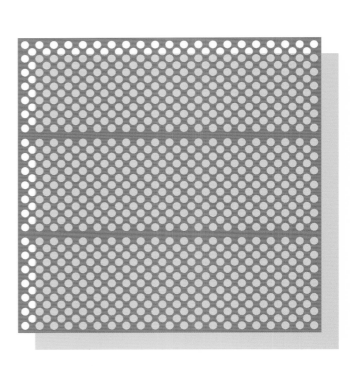

Kaleidoscope colour study

Facade details

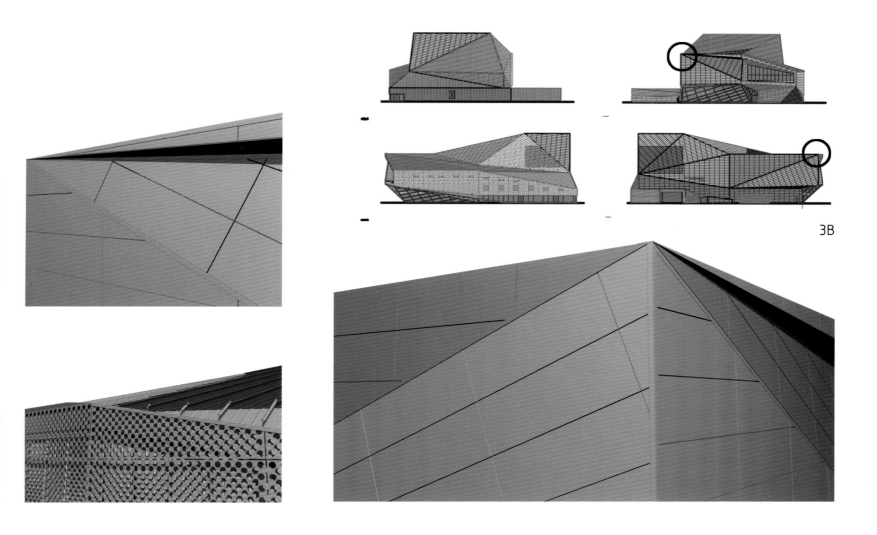

3B

A passtuk goot-goot-verbinding

B afsnijding 3 vlakken; komen in niet in een punt uit

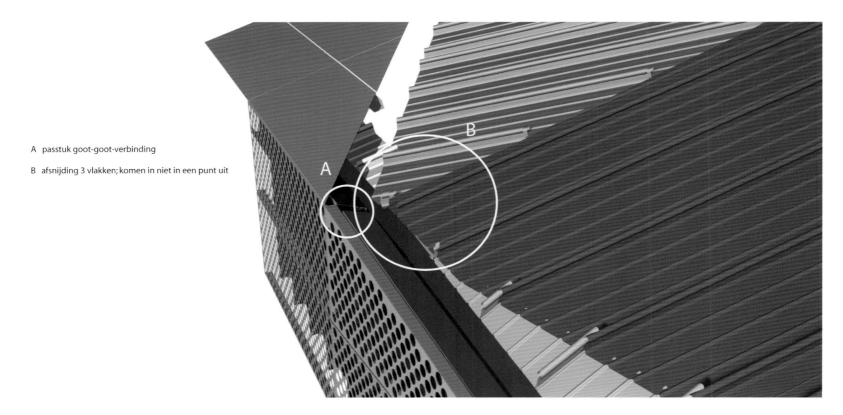

Connection detail

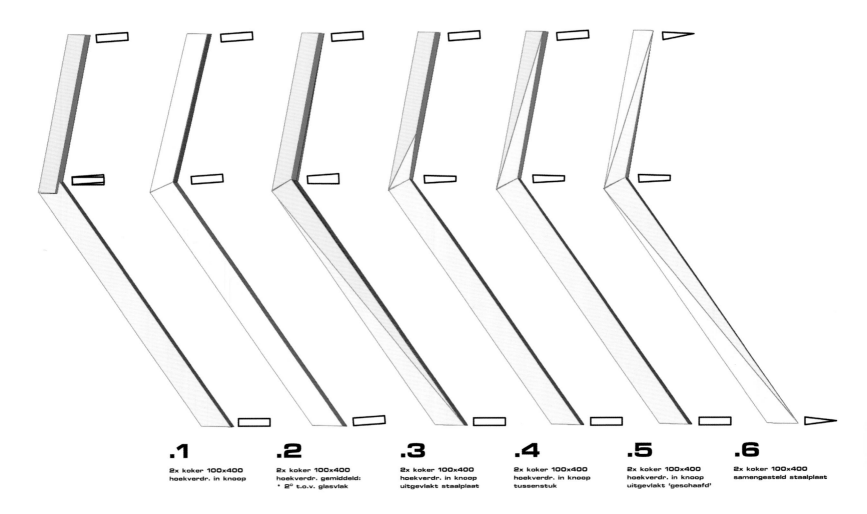

Study for corner solution, North facade

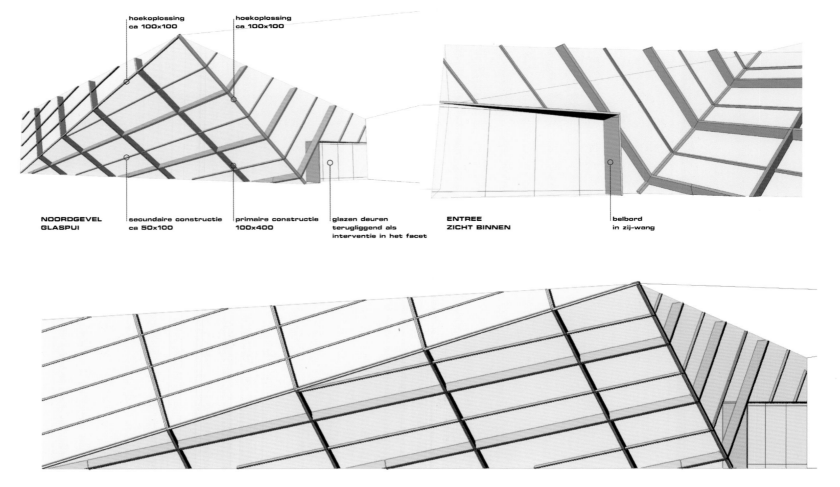

Study North facade / entrance

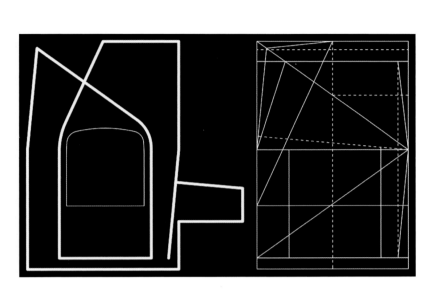
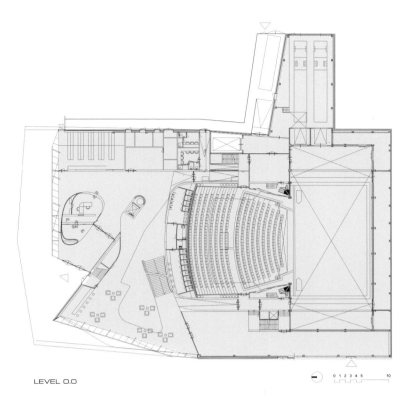

LEVEL 0.0

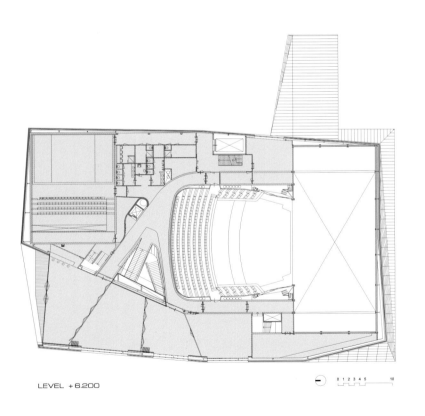

LEVEL +6.200

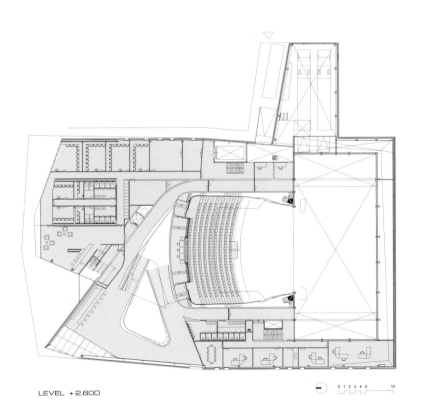

LEVEL +2.600

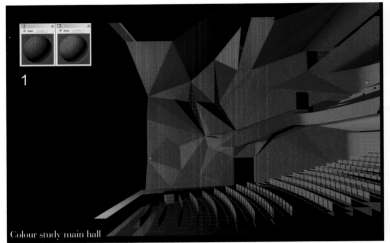
Colour study main hall

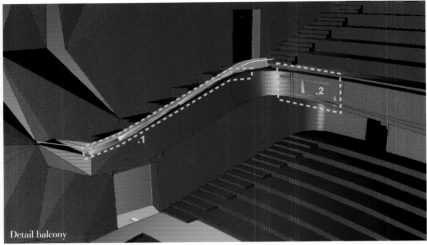
Detail balcony

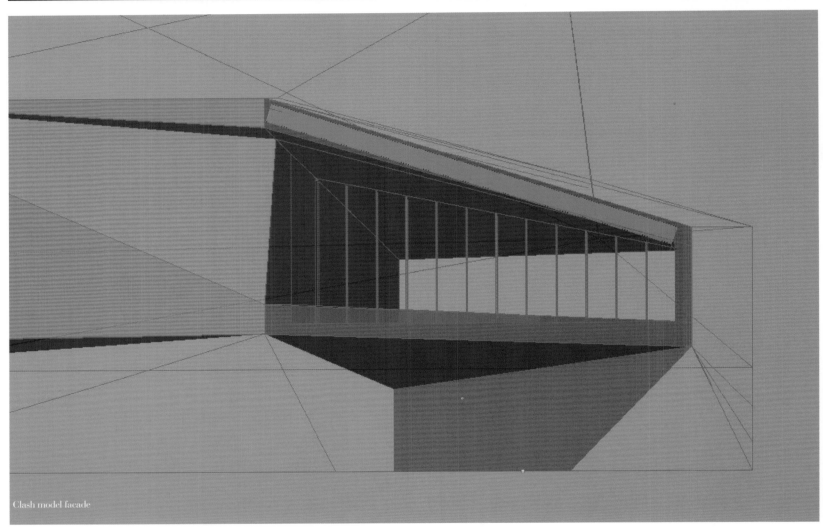
Clash model facade

  Agora 剧院是一个色彩斑斓、活力十足的地方。这座建筑是 Adriaan Geuze 为莱利斯塔德市制作的总体规划设计中的一部分，旨在复兴这个务实、宁静的城镇中心。这座剧院承载着战后荷兰小镇复兴的使命，专注于剧院的原始功能：创造一个既有迷惑性又有魅力的世界。剧院内外墙上都有很多的切面，重现了舞台上千变万化的经历，让人无法分辨出梦幻与现实。在 Agora 剧院里，戏剧和表演并不仅仅局限于舞台与夜晚，更被延伸到了城市生活和白天。

  在剧院内，外界的精彩正在上演；栏杆扶手像一条舞动的粉红丝带般倾泻到主楼梯处，在底楼大而开阔的大厅中心恣意地蜿蜒，然后沿着墙向上延伸到屋顶，途中不断地变换颜色，由紫罗兰色变换到深红色，再到樱桃红直到几乎完全转变成白色。

  主剧院全部都是红色，舞台很大，能够承办盛大的国际演出，这对于如此小的城镇来说是很罕见的。马蹄形的剧院楼厅和护音墙充满生机的形式和阴影效果都提升了礼堂的亲民度。

287

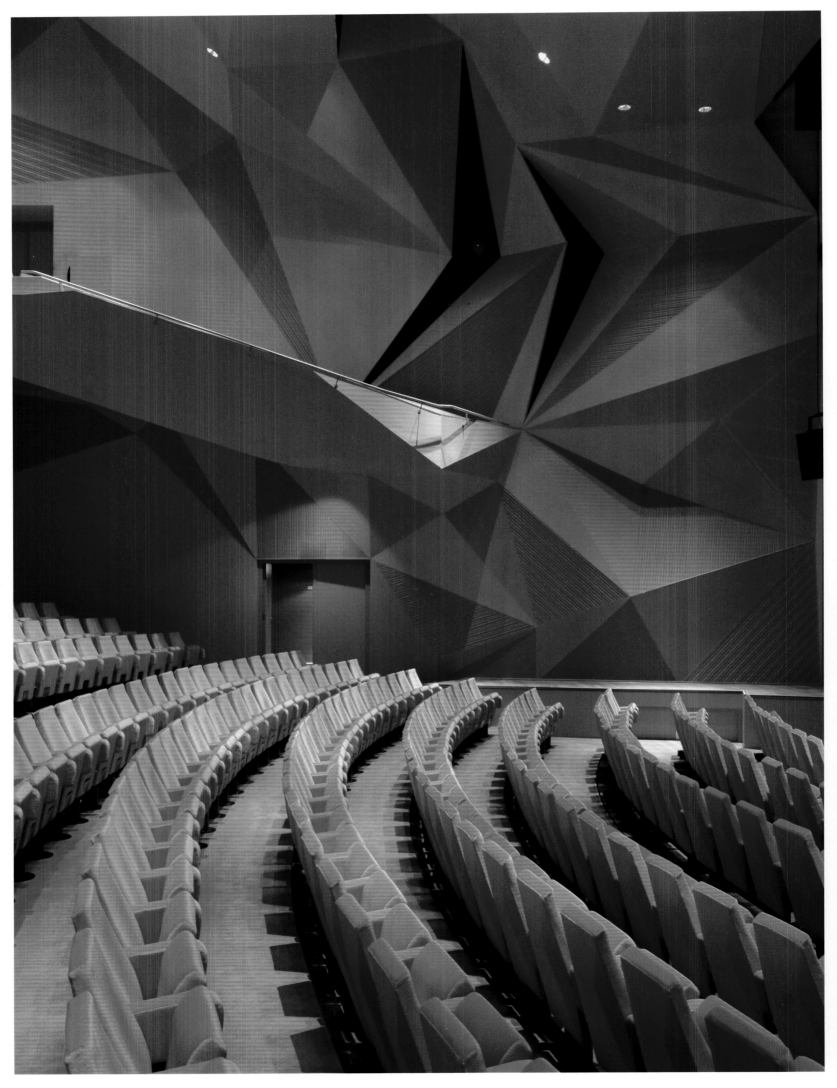

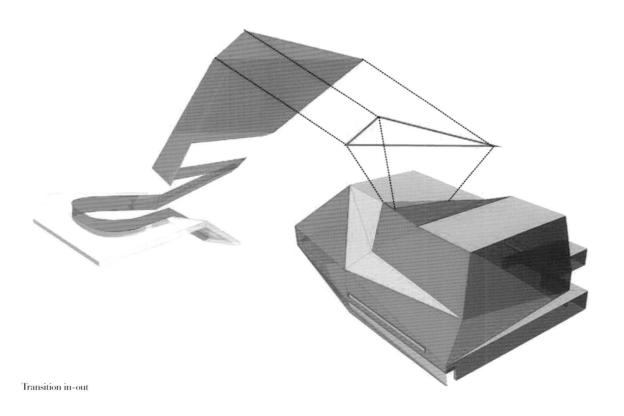

Transition in-out

LOCATION_Ireland, UK

WEXFORD OPERA HOUSE

Architect_Keith Williams Architects with the Office of Public Works
Area_7,235m²
Photographer_Ros Kavanagh

Wexford Opera Festival is as important culturally to Ireland as the Glyndebourne Festival is to England, consequently the building of the new 7,235 sqm Festival Opera House by Keith Williams Architects and the Irish Government's (Office of Public Works architects department) was Ireland's most important arts project for many years. The new opera house has been constructed in the heart of the medieval maritime town, on the site of the Festival's former theatre. It contains the new main opera house (780 seats) completely lined in North American black walnut, full flytower and backstage and a transformable second space of 175 seats, together with rehearsal, production facilities, bars, café and foyer spaces. The award winning building was officially opened by Mr Brian Cowen TD An Taoiseach (Irish Prime Minister) on 5 September 2008, whilst the first opera took place in the new house on 16 October 2008 with a performance of Rimsky-Korsakoff's Snegourchka (the Snow Maiden).

• Site Plan

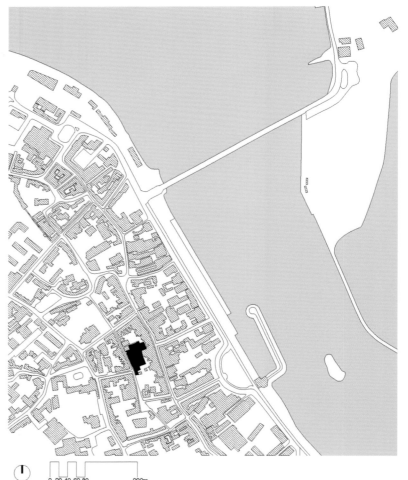

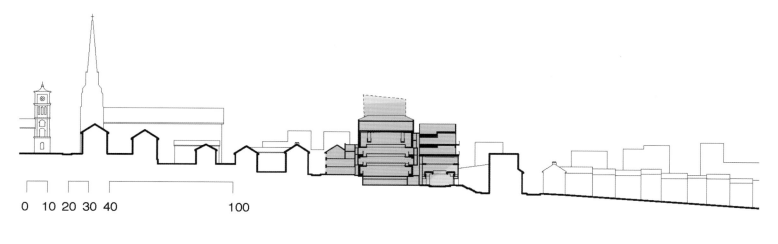

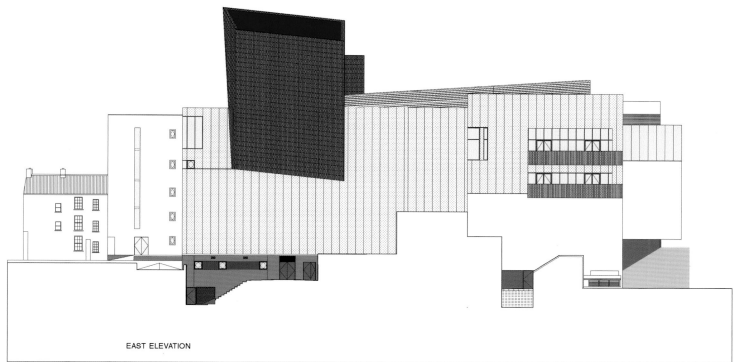

EAST ELEVATION

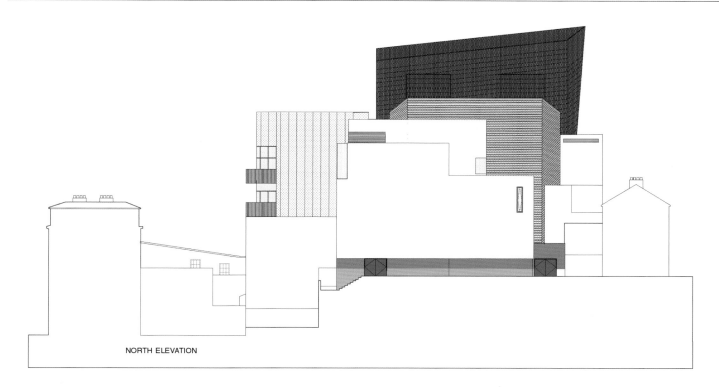

NORTH ELEVATION

295

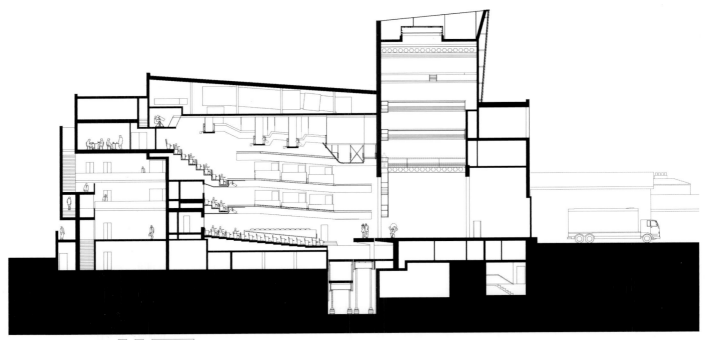

• Section A-A

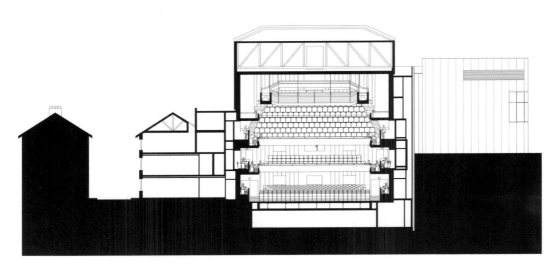

• Section F-F

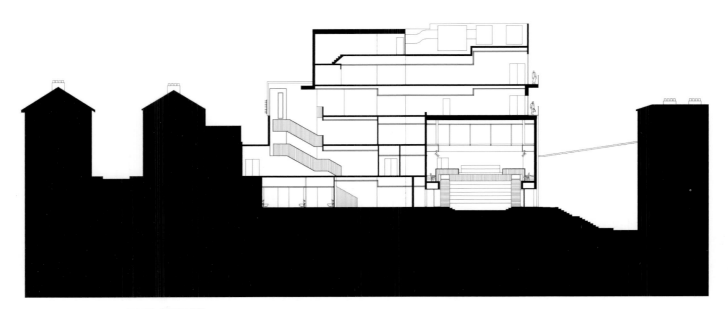

• Section H-H

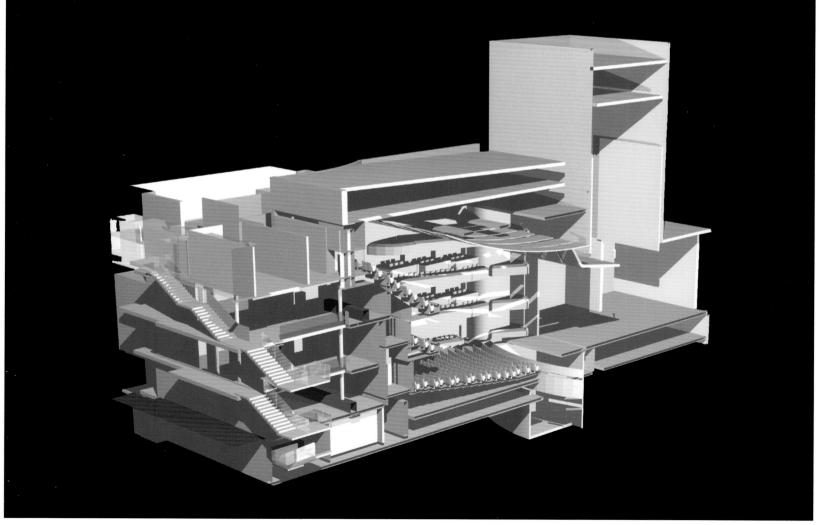

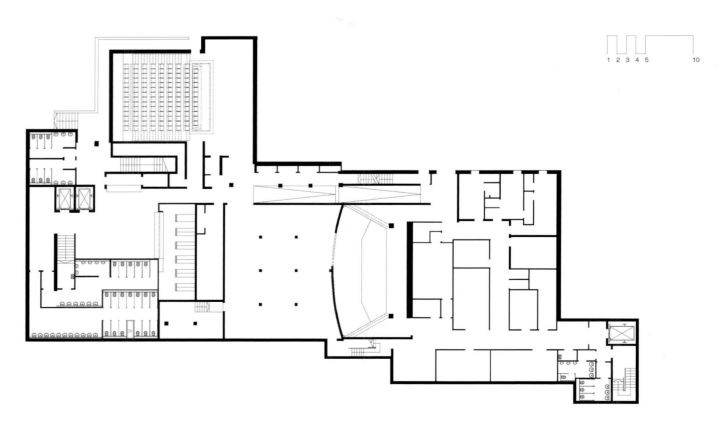

LEVEL -1

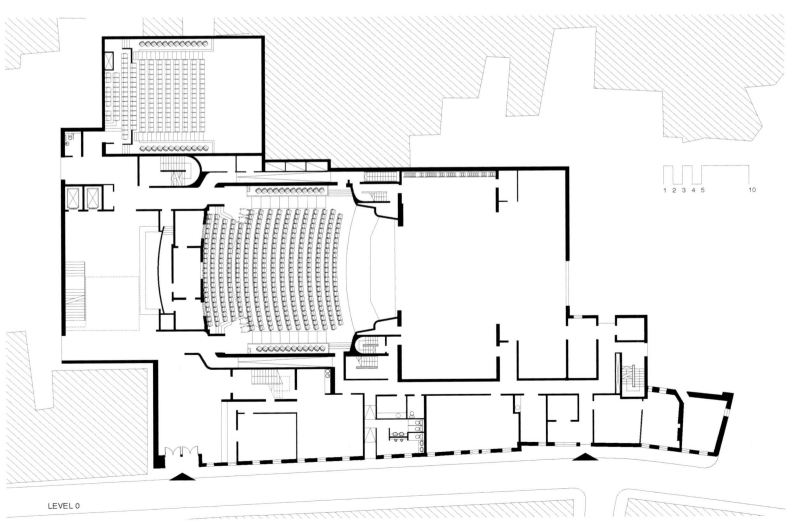

LEVEL 0

韦克斯福德歌剧节是爱尔兰重要的文化节日，就像英国的格林德伯恩艺术节一样重要。因此，这座占地面积 7235 平方米，由 Keith Williams 建筑师事务所和爱尔兰政府（公共建筑部门）合作完成的新节日剧院多年以来一直是爱尔兰最重要的艺术建筑。新的歌剧院设立在中世纪风格的海边小镇的中心、原节日庆祝剧院的旧址上。这个剧院包括陈列在北美黑胡桃质地板上的新主歌剧院（内设 780 个座位）、舞台塔、后台及拥有 175 个座位的过渡区，此外还有排练室、设施区、酒吧、咖啡厅和休息厅。2008 年 9 月 5 日这座获奖的建筑由爱尔兰国家议会众议院的 Brian Cowen（爱尔兰总理）正式宣布对外开放。在2008 年 10 月 16 日剧院举行了首场歌剧演出——里姆斯基·柯萨科夫的 Snegourchka（雪姑娘）。

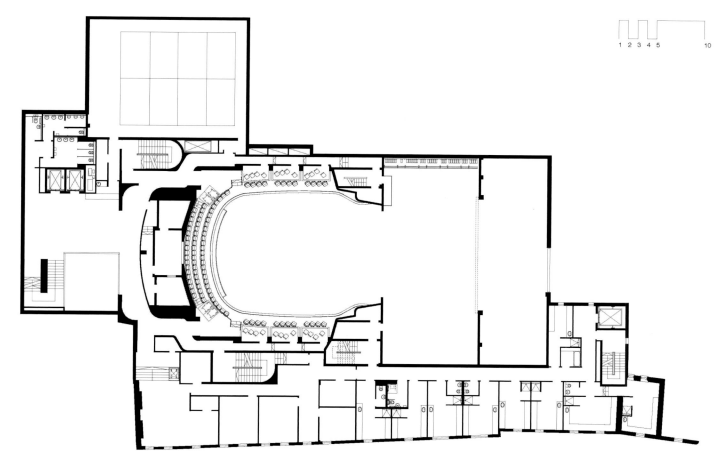

LEVEL 1

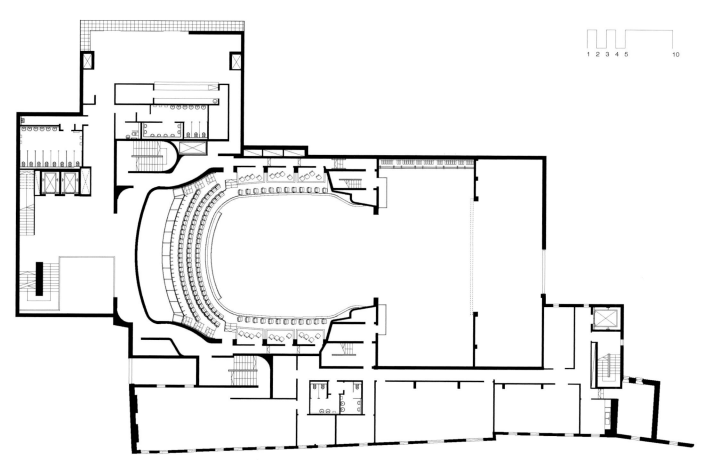

LEVEL 2

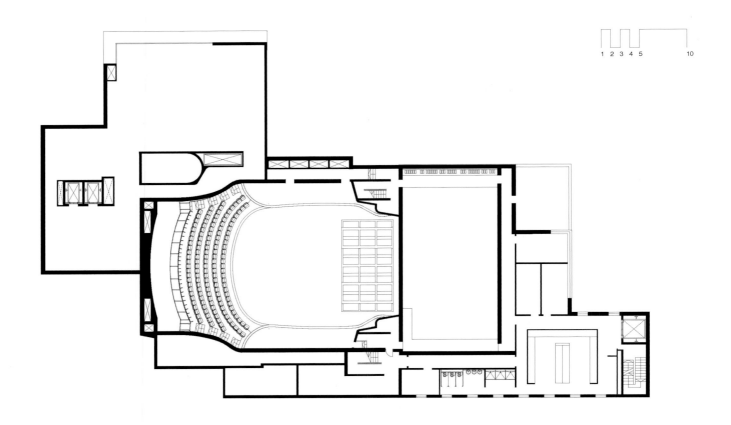

LEVEL 3

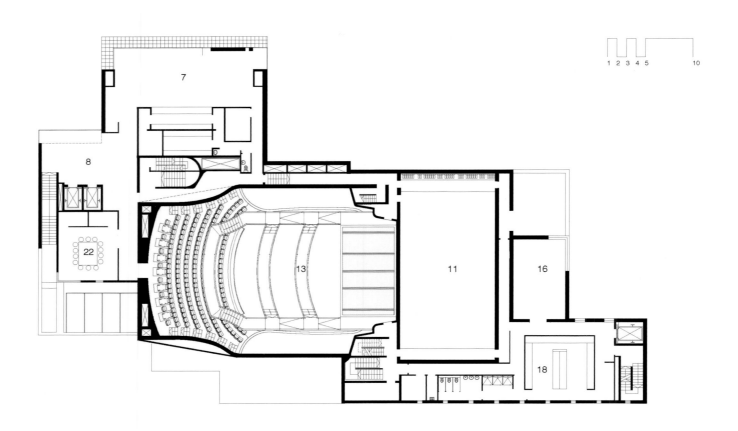

LEVEL 4

LOCATION_ Geelong, Victoria, Australia

THE PLAYHOUSE THEATRE

Design Company_studio101 architects
Area_650m²
Photographer_Trevor Mein (meinphoto)

The refurbishment of the 800-seat proscenium arch theatre "The Playhouse" at Geelong Performing Arts Centre is a landmark project for the regional city of Geelong. The underlying concept for the design was heavily forged through the innovative use of timber and has vastly improved the spatial quality and acoustics from the original theatre opened some 30 years ago. Universal access requirements, patron seating comfort, lighting and AV equipment were also improved and upgraded. Inspired by the nearby Otway Ranges State Forest, the main palette for the refurbishment consists of natural timbers and luscious green fabrics. The remaining existing surfaces inside the theatre have been painted black to become recessive planes, allowing the Walnut timber forms to seemingly push forward and interact with the large spatial volume. This hierarchy of materiality was enhanced with the strategic use of LED timber pelmet lighting that washes and highlights the timber linings with a beautiful even glow. The Playhouse redevelopment has significantly enhanced the Geelong cultural precinct's public profile with the consistent expression, detail and underlying conceptual use of materials transforming the once dated theatre into a finely crafted performance space that marks a centrepiece to Geelong's dynamic arts hub.

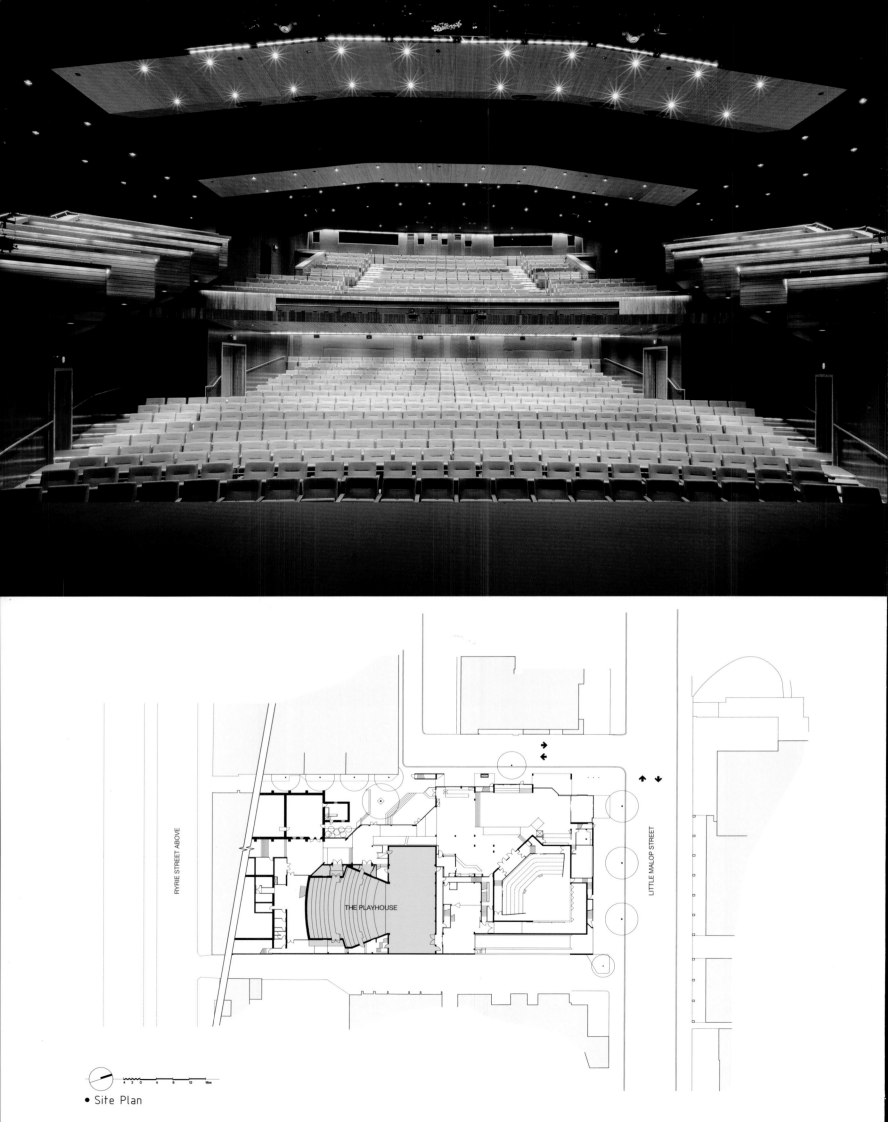

• Site Plan

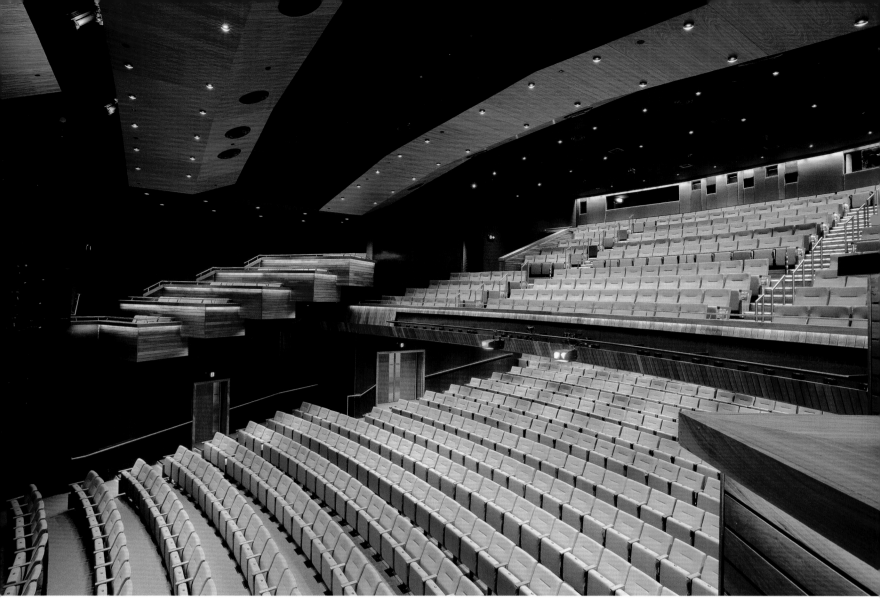

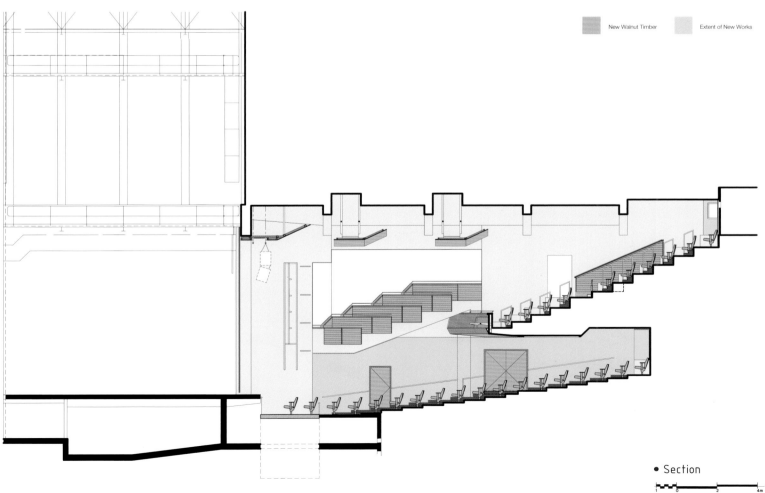

New Walnut Timber   Extent of New Works

• Section

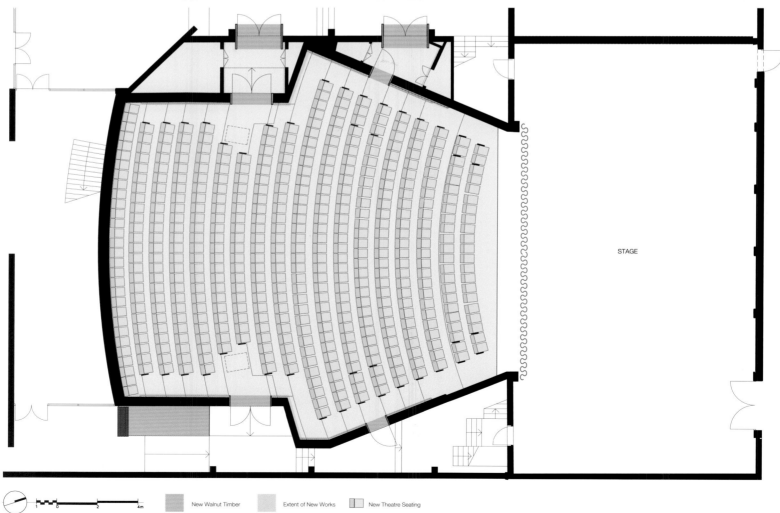

• Stalls Floor Plan

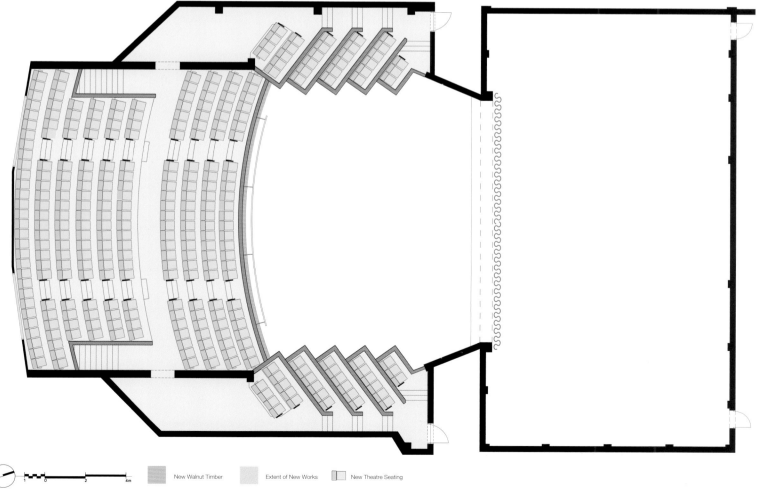

New Walnut Timber   Extent of New Works   New Theatre Seating

• Balcony Floor Plan

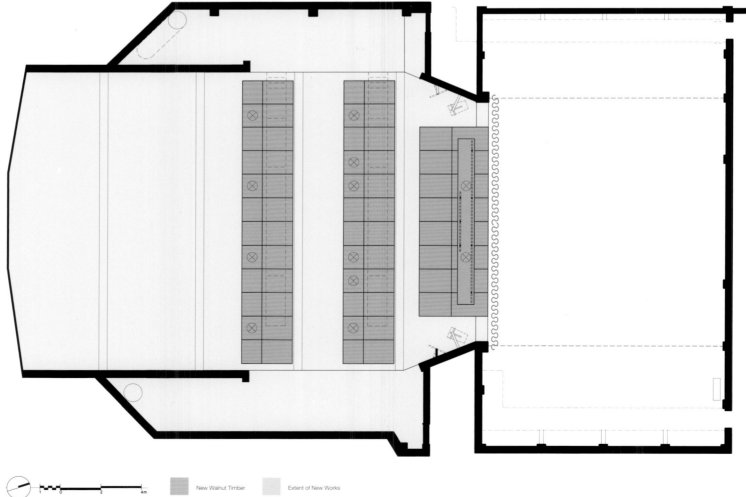

New Walnut Timber   Extent of New Works

• Balcony Reflected Ceiling Plan

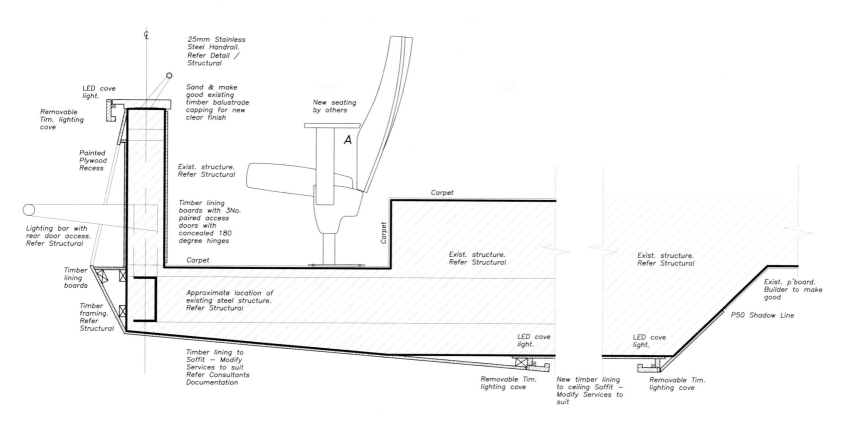

• Balcony Detail

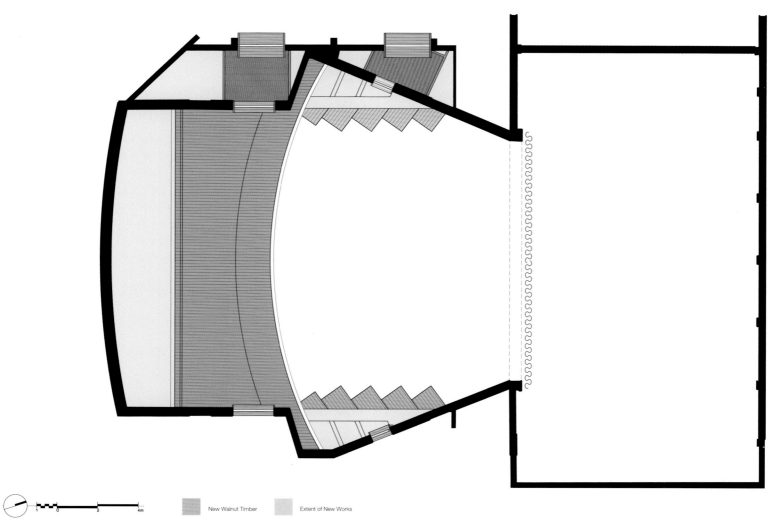

• Stalls Reflected Ceiling Plan

311

季隆艺术表演中心是季隆市的标志性建筑。这座拥有800个座位的拱形舞台剧院翻新的基本理念为创新地使用木材，对空间品质及30年前的原始剧院的音效进行了大规模的改进。统一了进馆要求，同时也对座位舒适度进行了改进并更新了灯光及视听设备。受附近的奥特威国家森林区的启发，翻修的主要材料采用了天然的木材和绿色纤维织物。剧院内部其余现有表面均被涂成黑色，形成视觉上向后的平面，这使得胡桃木材质部分看起来似乎向前推进了，与大面积的空间相呼应。木质窗帘盒内LED射灯的巧妙运用极大地增强了材料的层次感，美丽柔和的灯光徜徉于木衬之上使其成为焦点。剧院的改造极大地提升了季隆文化区的公共形象，一如既往地宣扬细节和基本材料的使用理念，将曾经过时的剧院改造成为雅致精美的表演场地，成为季隆最具活力的艺术中心的一大亮点。

CONCERT HALL

音乐厅

Situated on the border between land and sea, the Concert Hall stands out as a large, radiant sculpture reflecting both sky and harbor space as well as the vibrant life of the city. The spectacular facades have been designed in close collaboration between Henning Larsen Architects, the Danish-Icelandic artist Olafur Eliasson and the engineering companies Ramboll and ArtEngineering GmbH from Germany.

The Concert Hall of 28,000m^2 is situated in a solitary spot with a clear view of the enormous sea and the mountains surrounding Reykjavik. The building features an arrival and foyer area in the front of the building, four halls in the middle and a backstage area with offices, administration, rehearsal hall and changing room in the back of the building. The three large halls are placed next to each other with public access on the south side and backstage access from the north. The fourth floor is a multifunctional hall with room for more intimate shows and banquets.

Seen from the foyer, the halls form a mountain-like massif that similar to basalt rock on the coast forms a stark contrast to the expressive and open facade. At the core of the rock, the largest hall of the building, the main concert hall, reveals its interior as a red-hot centre of force.

The project is designed in collaboration with the local architectural company, Batteríið Architects.

LOCATION_ lReykjavik, Iceland

HARPA - REYKJAVIK CONCERT HALL AND CONFERENCE CENTRE

Architect_ Henning Larsen Architects and Batteriid Architects

Drawings_ Henning Larsen Architects

Area_ 28,000m^2

Photographer_ Nic Lehoux

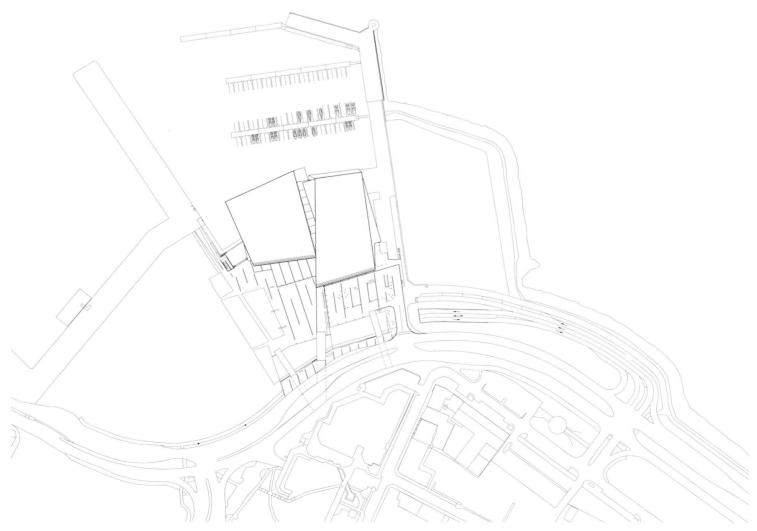

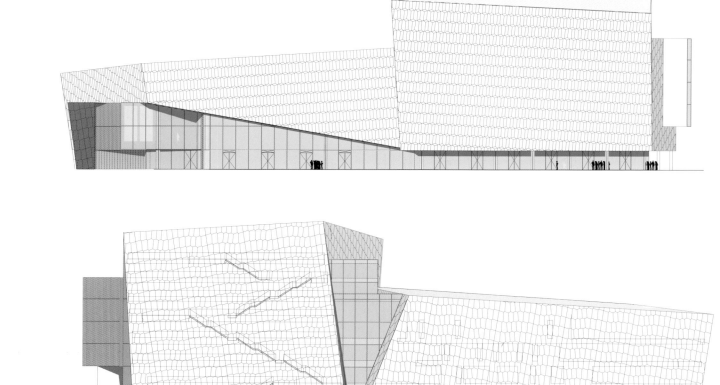

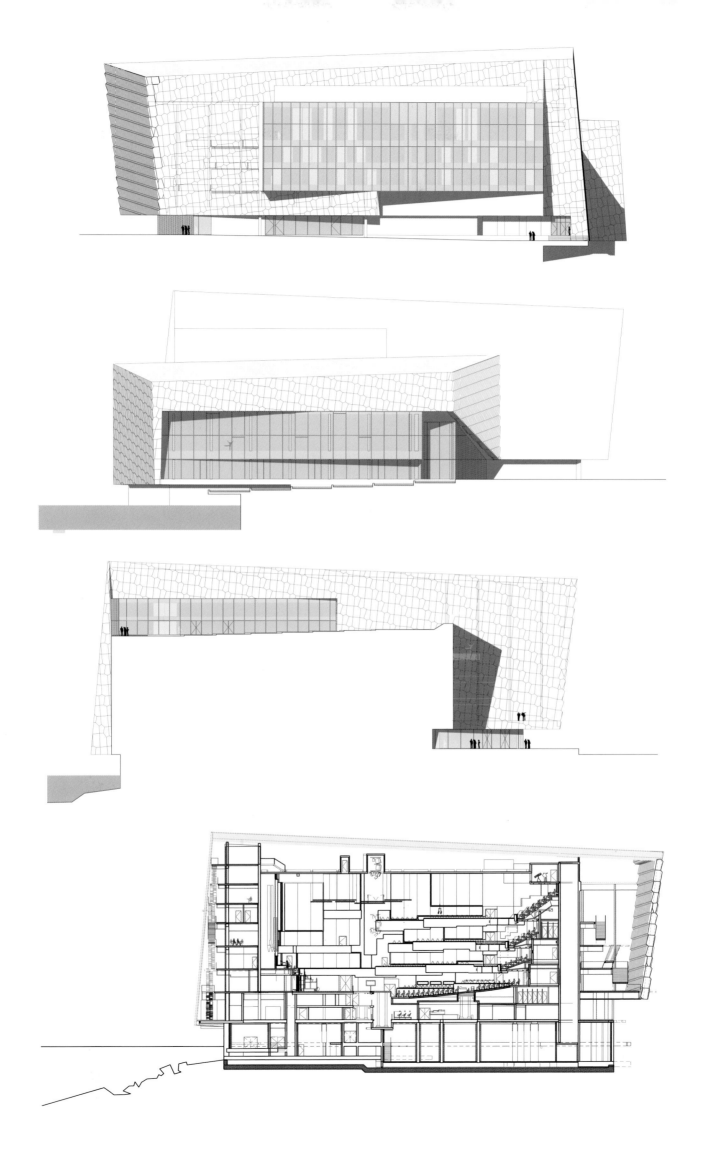

321

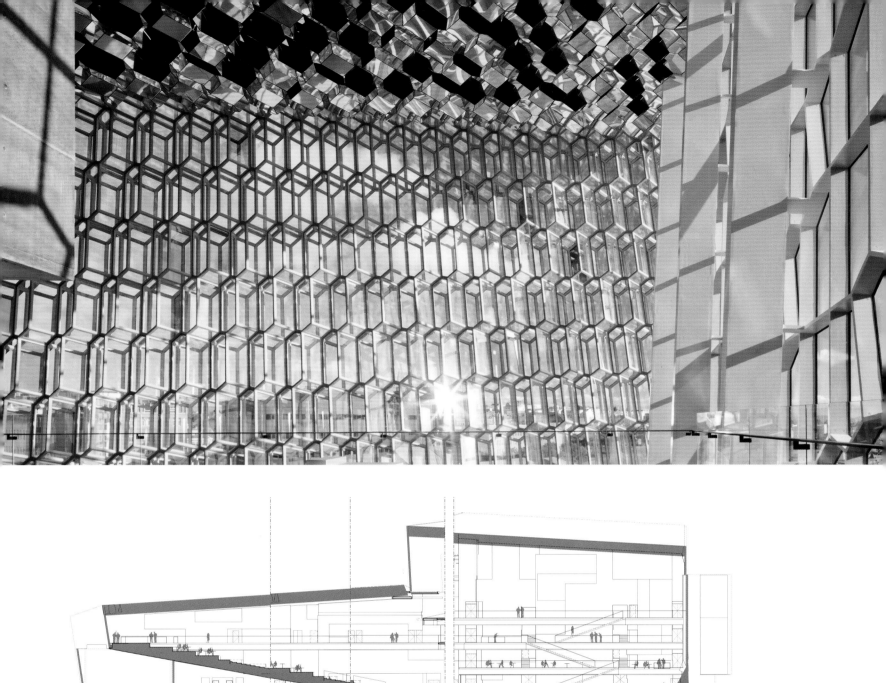
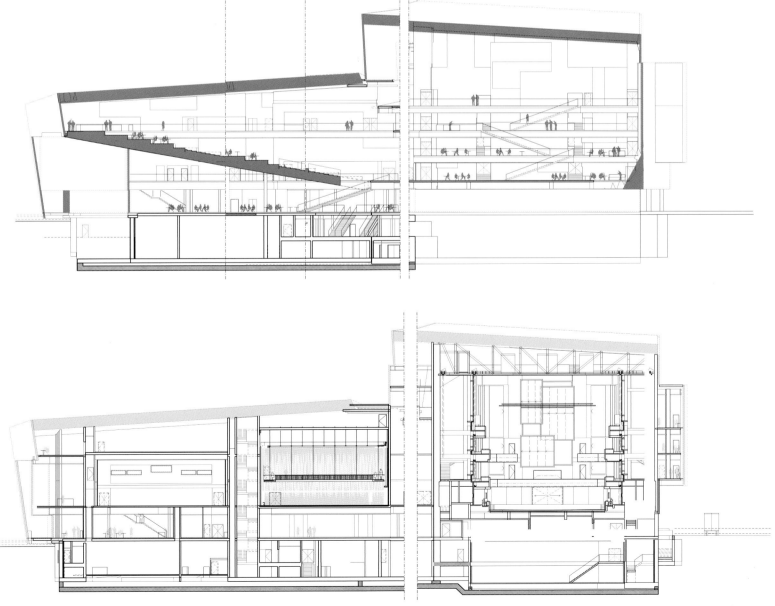

323

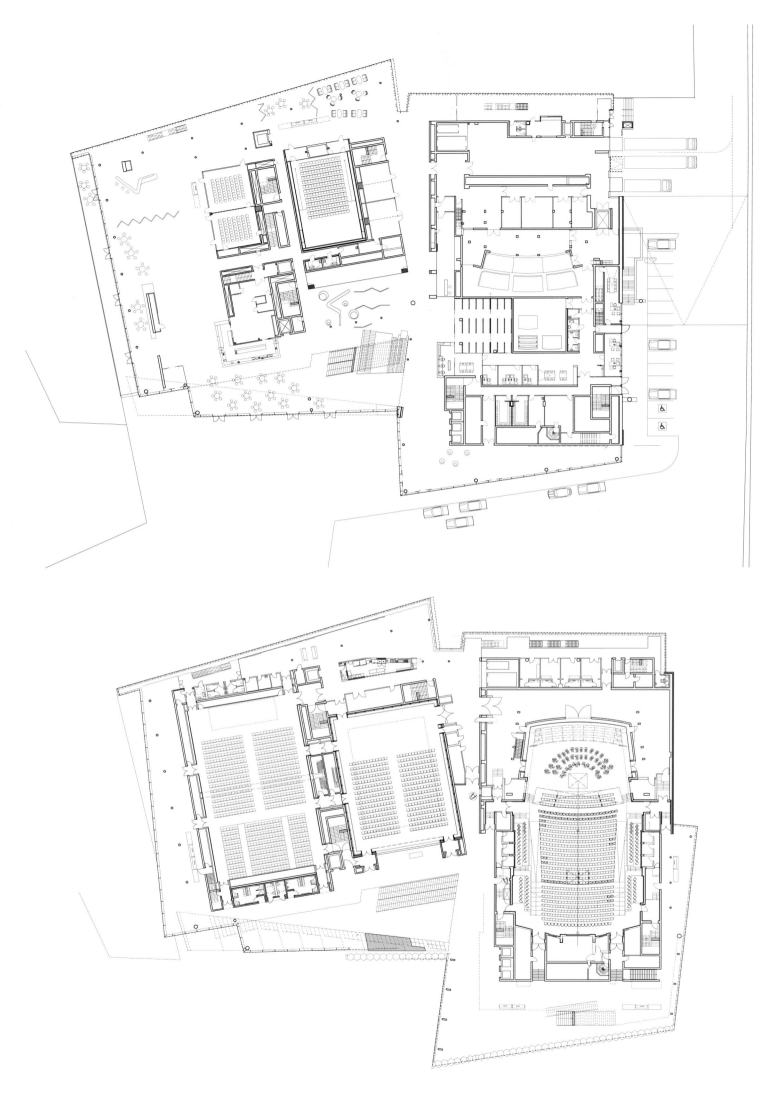

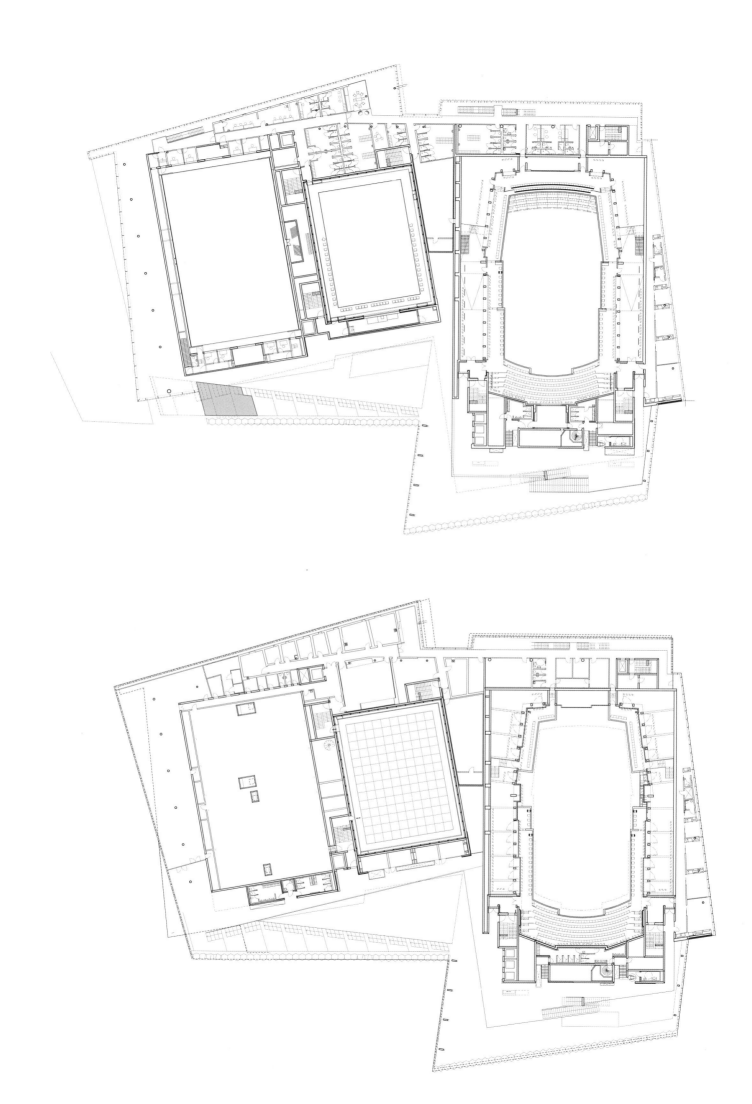

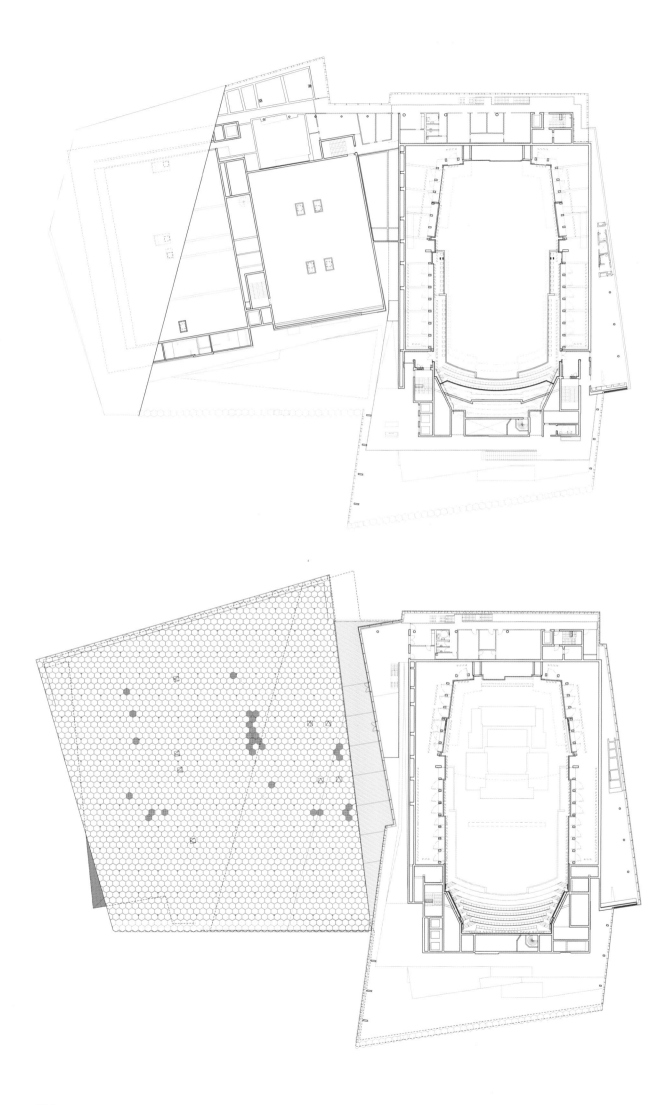

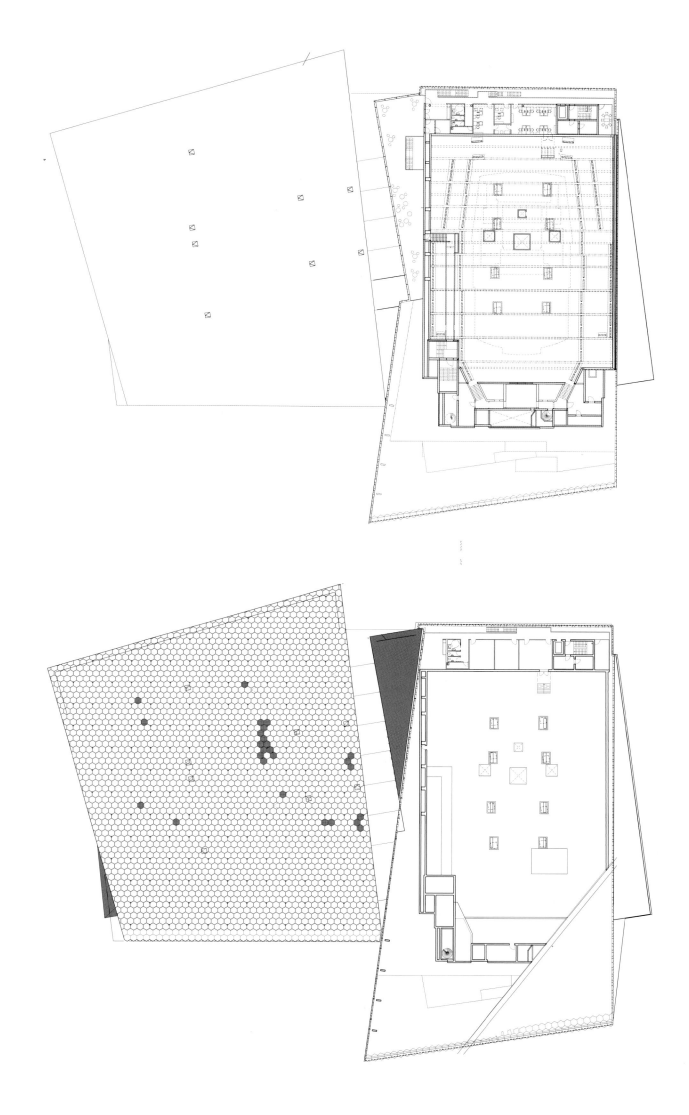

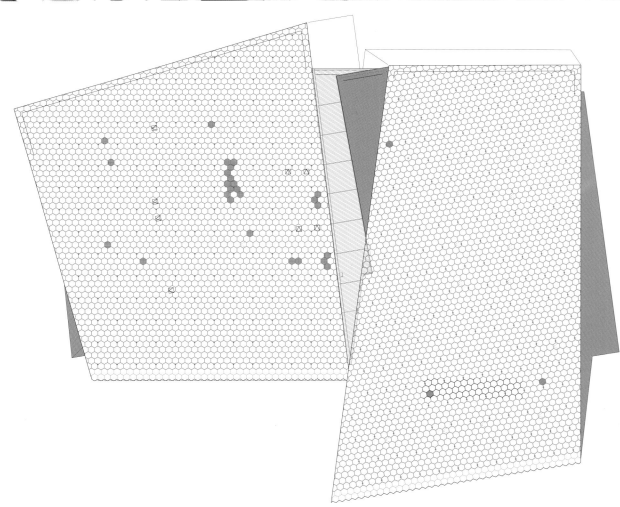

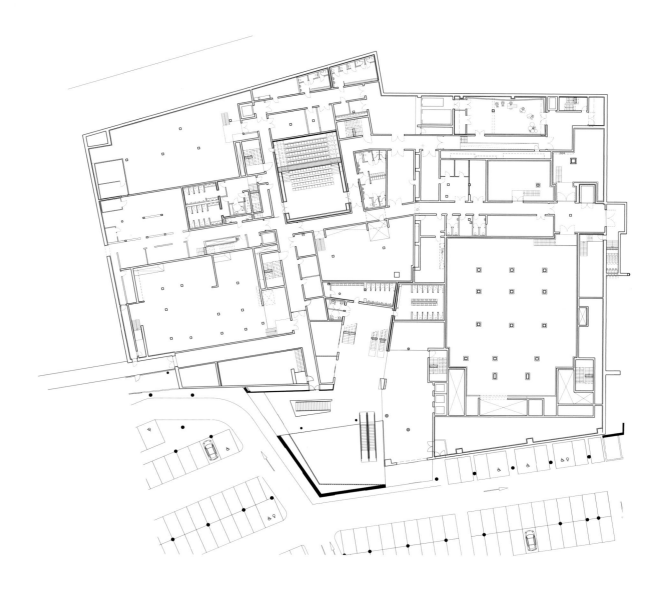

音乐厅坐落于海陆交界处,其巨大而光芒四射的雕塑形体映照着天空与海港,同时也展现了这里活力四射的城市生活。音乐厅壮丽的外观是由Henning Larsen建筑师事务所、丹麦籍冰岛艺术家Olafur Eliasson、德国建筑公司Ramboll及艺术工程公司GmbH密切合作完成的。

28000平方米的音乐厅位于一个独立场所,从那里可以清晰地观赏到围绕在雷克雅未克周围的浩瀚海洋与连绵群山。建筑的前方区域是一个抵达接待区,中间区域有四个大厅,建筑的后方区域是后台区,设有办公室、行政处、彩排大厅和更衣室。三个区域依次排列,南侧有公共通道,北侧为后台通道。四楼是多功能厅,用于举办私人表演及宴会。

从接待区看去,整个大厅状似一座褶皱山,宛如海岸边的玄武岩,与生动奔放的建筑外形形成鲜明对比。"玄武岩"的核心之处即这栋建筑中最大的大厅,主音乐厅,是一个洋溢着红色热情的室内空间。

这个设计项目是在当地建筑公司Batteríð建筑公司的协助下完成的。

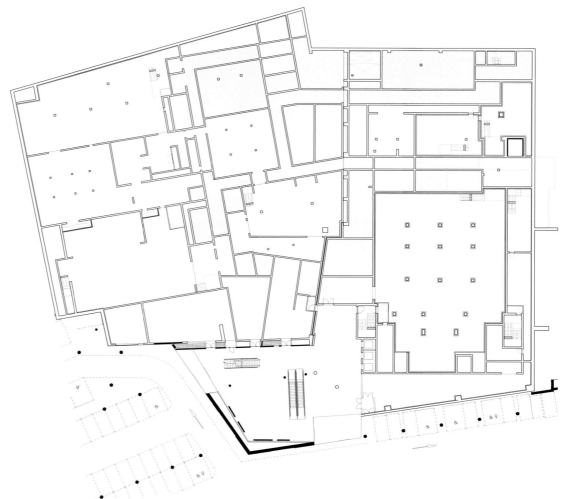

LOCATION_Eindhoven, the Netherlands

MUSIC HALL EINDHOVEN

Designer_Van Eijk & Van der Lubbe

Area_11,000m^2

Photographer_Frank Tielemans

Designers Van Eijk & Van der Lubbe have designed a concert hall that is nothing short of futuristic, complete with lighting, design and technology fully tailored to visitors. Since the official opening, thousands of interested parties have visited the new hall. In the middles of Eindhoven now stands the absolute music centre of the future, a place where lighting, design and technology are integrated innovatively, without it becoming merely a high-tech building. Niels van Eijik and Miriam van der Lubbe of Geldrop designed both the interior and the exterior around the central idea of the Concert Hall as a meeting place. Take for example the way concert-goers are led intuitively from foyer to concert hall by way of subtle lighting signs which move from a high-tech wall and over the ceiling. Van Eijk & Van der Lubbe of Geldrop designed every aspect of the building especially for the concert hall, from the gigantic glass facade to the foyers, and from the uniforms to the dishes and furnishings. This includes innovative listening chairs, where visitors can sit and listen to music in peace and quiet as well as love seats in an intimate section of the foyer with lighting that automatically dims when someone sits in them. The most remarkable change is to the main entrance of the new Concert Hall. This consists of a forward-leaning glass facade, 25 meters wide and 13 meters high. Behind it is a cultural city-foyer where people are welcome throughout the day for a cup of coffee, and to listen to, and buy music. The city-foyer will be fitted with an ambient wall, several meters long, consisting of thousands of led lights, on which films, works of art and concerts will be projected. Visual artist Gerald Hadders realised the content of this living wall.

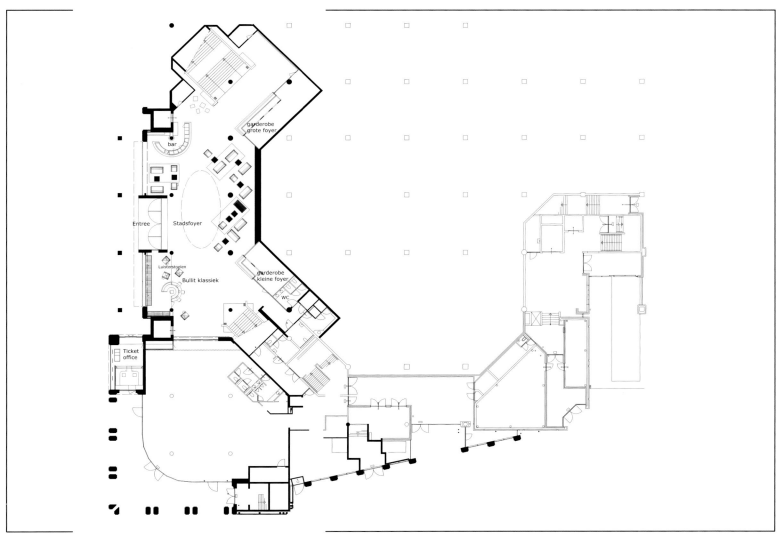

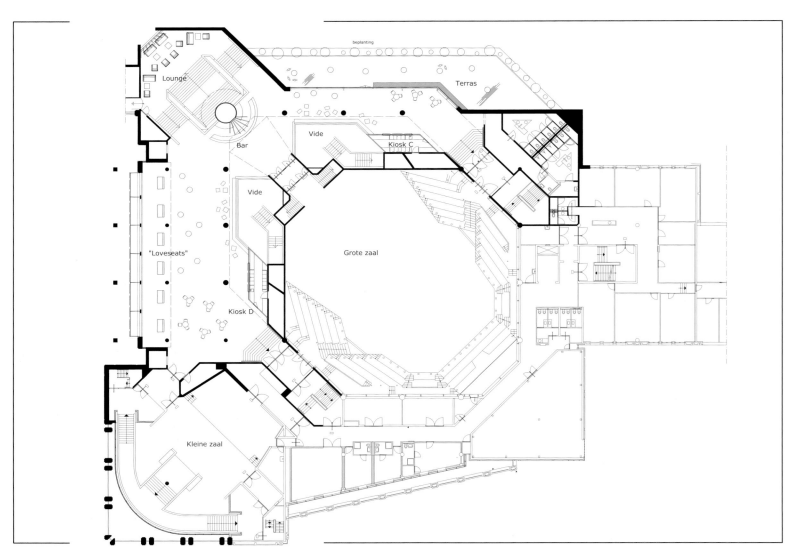

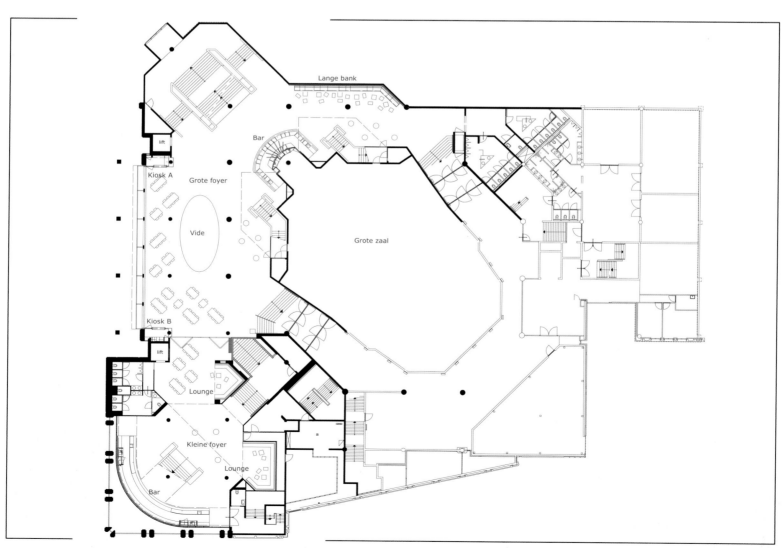

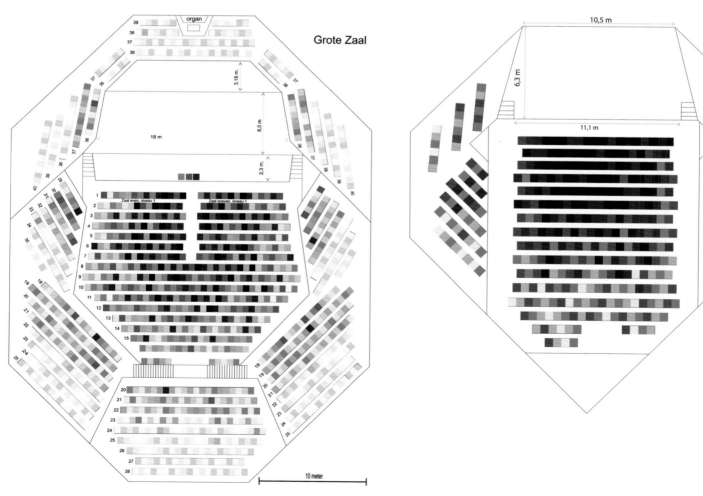

Grote Zaal

设计师 Van Eijk 和 Van der Lubbe 设计的音乐厅富有未来感，灯光、设计、技术完全为观众量身打造。自从正式开放后，已有数以万计对此感兴趣的人士参观了这座新建的音乐厅。矗立于埃因霍温市中心的这座音乐厅绝对会是未来的音乐中心，这里将灯光、设计与科技完美结合，独特而新颖，而若缺少了这些元素它只能被称为一座高科技建筑。来自荷兰 Geldrop 的 Niels van Eijk 和 Miriam van der Lubbe 设计师围绕着将音乐厅设计成会议地点的中心思想对其进行了内部和外部设计。比如，灯光标志发出的柔和的光从高科技墙面一直射到天花板上，直观地引领着音乐会的观众们从门厅进入音乐厅。从巨大的玻璃幕墙到门厅，从制服到餐具及家具，两位设计师对建筑各个方面尤其是音乐厅都进行了精心的设计，这其中还包括创新的听椅，观众可以坐下来静静地聆听音乐；同样，门厅内还设有亲密的情侣座椅，客人落座后光线会自动调到微暗色调。新音乐厅最醒目的改动之处便是音乐厅的主要入口处，设置了宽 25 米、高 13 米的前倾玻璃幕墙。在它的后面是一个城市文化厅，人们在参观了一天的音乐厅后可以在这里坐下喝杯咖啡、听听音乐或者买张唱片。城市文化厅几米长的外围墙壁上将内设数以千计的 LED 灯，电影、艺术及音乐作品将在上面播放。视觉艺术家 Gerald Hadders 充分展现了这面墙栩栩如生的一面。

LOCATION_Poitiers, France

TAP . THEATRE AUDITORIUM OF POITIERS, FRANCE

Architect_João Luís Carrilho da Graça

Structural Engineering_DL Structures

Landscape architecture_GLOBAL - João Gomes da Silva, architect

Graphic Design_P-06 ATELIER, Nuno Gusmão, designer

Area_32,000m^2

Photographer_FG+SG Fotografia de Arquitectura

Deciding what is essential in a given programme for a given place should be the primary objective of every single architecture project – and nothing else. The limestone platform open to the public ensures a spatial continuity with the city and a material homogeneity with the surroundings. Slightly suspended above it lie the parallelepiped volumes of the building, covered with white matt glass, which works as a medium and allows mutations in the building exterior – of color, light, image…

The theatre

The theatre hall was meant to be extremely versatile and "performant", in order to allow different kinds of productions and events to take place. The theatre is not only a "theatre machine", but also a visual and acoustic optimization of audience space in a balanced and flexible way. The audience space is totally configured by gypsum fibre plates that produce a unitary shape, a dark, neutral, monochromatic "cocoon", only punctuated by the doors, control room and VIP galleries. Its homogeneous shape and materials ensure acoustic effectiveness, its simplicity emphasizes the stage performance.

The auditorium

Designing a hall exclusively dedicated to music contributed to an optimal acoustic and architectural result. The typological shape of the hall is that of a shoebox, a large rectangular space with a flat seating area. On the interior, leaning walls made of wood are detached from the container's surface, producing a unitary space that incorporates both stage and orchestra. Their slightly round shape, dictated by acoustics, and their bright texture create a structure that contrasts with the container's darker and more rigid forms, so the delicate textured surfaces diffusing the sound provide acoustic perfection and a feeling of sensory well-being for performers and audience alike.

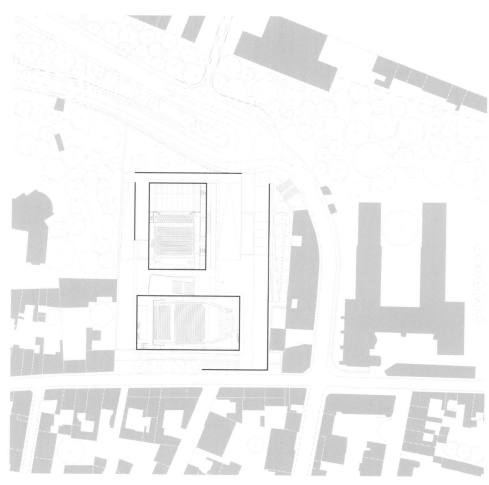

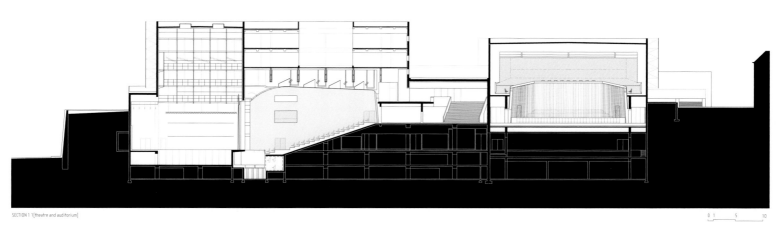

SECTION 1.1 [theatre and auditorium]

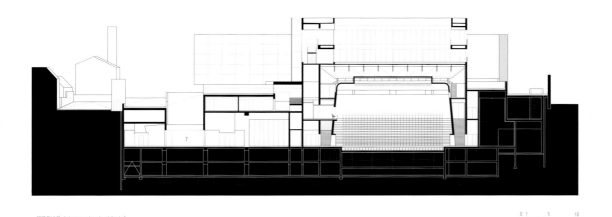

SECTION 2.2 [artists room and courtyard, theatre]

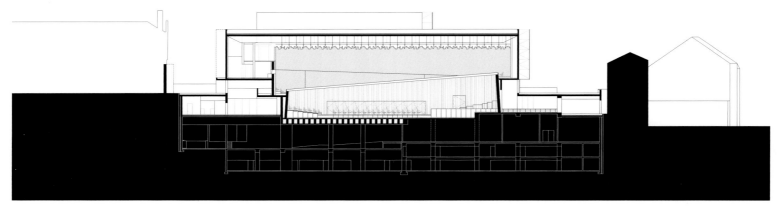

SECTION 3.3 [auditorium]

在特定地点特定项目中起到决定性作用的只能是建筑项目的主要目标。石灰石舞台向公众开放，以确保其在空间上与这座城市相连，在材料上与周围环境融为一体。表面为白色亚光玻璃的平行六面体的建筑部分略微悬于舞台之上，这个平行六面体建筑作为一种媒介使建筑外观的色彩、光、影都有所变化。

剧院

剧院大厅被设计成一个多用途高性能的所在，以适应不同类型的演出活动。这座剧院不仅是一个"用于表演的场所"，还是一个用平衡而灵活的方式最大限度地优化观众视听效果的场所。整个观众席的空间被纤维石膏板围成了单一的形状，成为了一个颜色深沉的、素雅的、单色的"茧"，只有通过门、控制室、贵宾廊才能划分出空间。它统一的外形和材料确保了音效，它朴实的设计突出了表演的效果。

礼堂

建造高级的音乐大厅需要最佳的声音效果和建筑设计。整个音乐厅是一个大型的矩形空间，状似鞋盒，厅内设有水平的座位区。室内，木质的墙体从墙面延伸出来，营造出一个与整体空间基调相同的场所，用作舞台及乐队表演。略显圆形的设计旨在优化音效效果，鲜艳的纹理与墙面暗色、庄重的外观形成对比。因此，刻有精美纹理的墙面可以扩散乐声进而创造出完美的音质，使演奏者与观众同时享受到美妙的音乐。

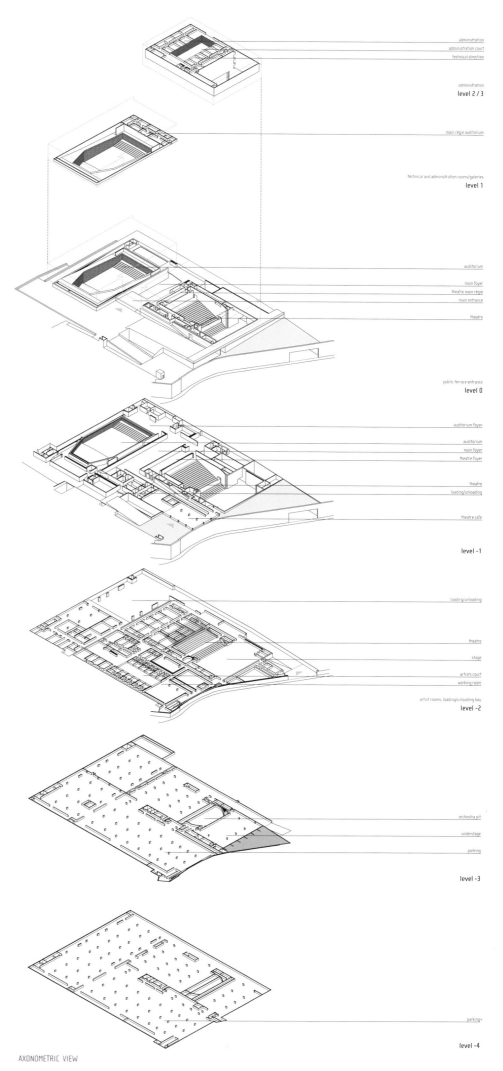

AXONOMETRIC VIEW

1. plasterboard with antivibratory suspending rods
2. 27 mm nordic pine planks over squared timber
3. plasterboard over galvanized steel structure
4. self levelling polyurethane
5. 20 mm calcareous stone

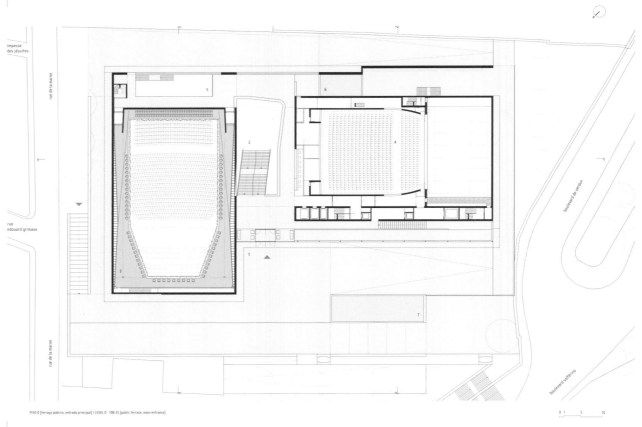

PISO 0 [terraço público, entrada principal] / LEVEL 0 108.35 [public terrace, main entrance]

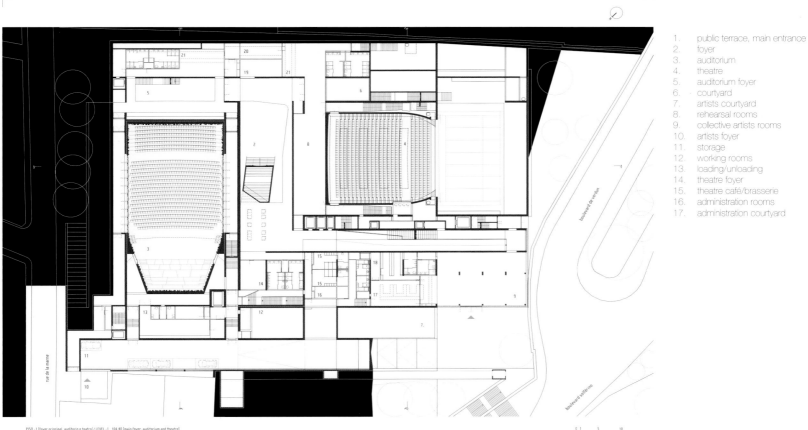

1. public terrace, main entrance
2. foyer
3. auditorium
4. theatre
5. auditorium foyer
6. courtyard
7. artists courtyard
8. rehearsal rooms
9. collective artists rooms
10. artists foyer
11. storage
12. working rooms
13. loading/unloading
14. theatre foyer
15. theatre café/brasserie
16. administration rooms
17. administration courtyard

LOCATION_Berkeley, CA

FREIGHT & SALVAGE

Design company_Marcy Wong Donn Logan Architects
Area_18,000m²
Material_Salvaged Wood Siding, Recycled Wood Trusses, Acoustic material, Green Roof
Photographer_Billy Hustace, Sharon Risedorph

West of the Mississippi, the Freight & Salvage is the country's most venerable institution dedicated to presenting folk and traditional music. "The Freight" became known for hosting quality performers in a welcoming, down-home environment. The Freight's new theatre space is designed to maintain and foster the venue's unique ambiance. The auditorium is designed with a level-floor stage to reduce separation between audience and performer, while deep rows of shallow-stepped seating allow optimal visibility without creating a "stadium effect". Movable seating means that rows can be reconfigured for table seating and standing audiences alike. Contemporary earth-toned seating complements the oil-stained Douglas Fir harvested during the demolition process and methodically repurposed as paneling for the auditorium interior. An absorptive acoustic backing means the stunning material is as effective sonically as it is visually. The theatre's façade and lobby area also maintains an entire 20'-deep(6.1m-deep) swath of the existing structure, and the architects were also able to reuse a majority of the old building's wood trusses. The use of wood in the project, and particularly salvaged wood, allowed the designers to create a striking aesthetic effect that would have been impossible to achieve with new construction. Furthermore, the extensive salvage operation provides an excellent example of how material reuse can reduce the environmental footprint of a building, keeping huge amounts of material out of landfills, and saving expense and energy that would otherwise go to replacement materials.

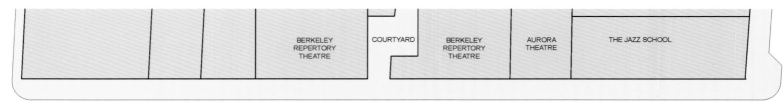

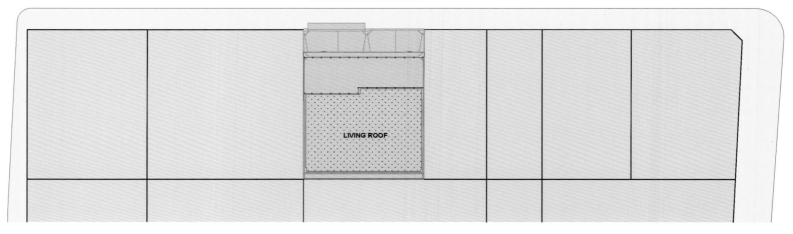

SITE PLAN AND PROJECT ROOF PLAN
0' 16' 32'

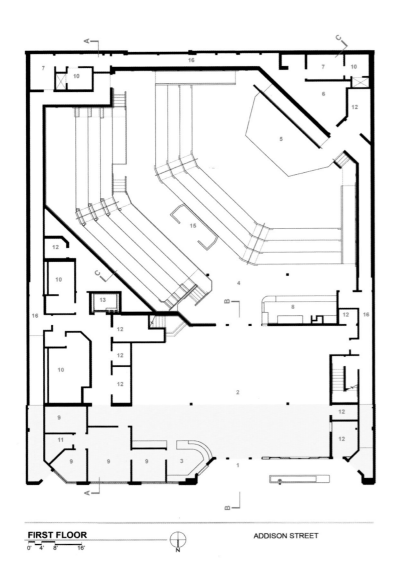

FIRST FLOOR

0' 4' 8' 16'

ADDISON STREET

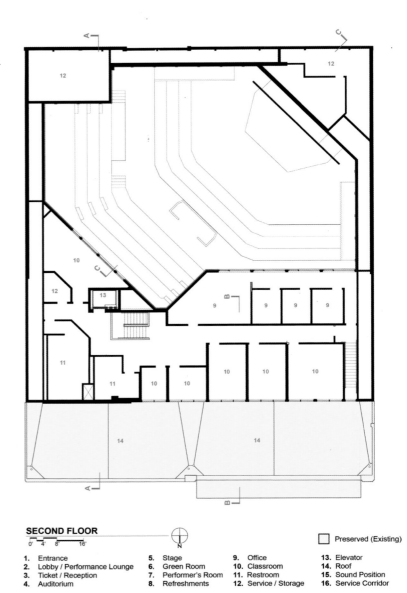

SECOND FLOOR

0' 4' 8' 16'

1. Entrance
2. Lobby / Performance Lounge
3. Ticket / Reception
4. Auditorium
5. Stage
6. Green Room
7. Performer's Room
8. Refreshments
9. Office
10. Classroom
11. Restroom
12. Service / Storage
13. Elevator
14. Roof
15. Sound Position
16. Service Corridor

☐ Preserved (Existing)

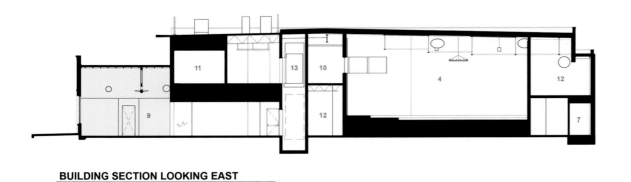

BUILDING SECTION LOOKING EAST

0' 4' 8' 16'

1. Entrance
2. Lobby / Performance Lounge
3. Ticket / Reception
4. Auditorium
5. Stage
6. Green Room
7. Performer's Room
8. Refreshments
9. Office
10. Classroom
11. Restroom
12. Service / Storage
13. Elevator
14. Roof
15. Sound Position

☐ Preserved (Existing)

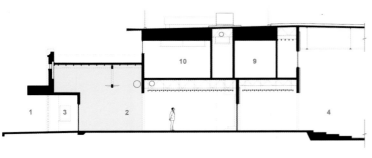

PARTIAL BUILDING SECTION LOOKING EAST

0' 4' 8' 16'

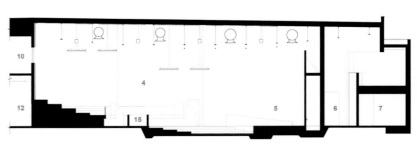

PARTIAL BUILDING SECTION LOOKING SOUTHEAST

0' 4' 8' 16'

358

密西西比河西岸的Freight & Salvage是一个备受人们喜爱的组织，主要致力于向人们展示传统民间歌谣和传统音乐。Freight因以热情的态度、如家的环境举行精彩的演出而闻名。Freight新剧院的设计很好地保留了场馆的独特气氛。观众席与舞台设计在一个高度上以缩短观众与演奏者之间的距离。一排排的低台阶式的座位呈现给观众最好的视觉感受，从而避免了"露天体育场"的效果。可移动座椅可根据每行桌子及站立观众的喜好进行重新配置。现代土褐色的座椅与拆迁时保留下来的油亮的花旗松木被有序地搭配在一起用作观众席的内部镶板，吸音效果与其视觉效果一样惊人。剧院的外观和大厅区域保留了20英尺（6.1米）深的原结构，建筑师们对旧建筑物的木制框架也进行了再利用。在整个项目中对木材的使用，尤其是废旧木材的使用，使设计者创造出一种新材料不可能创造出来的令人惊叹的审美效果。此外，大量的废物回收利用为物质的循环使用以及减少对环境的破坏做了一个良好的示范，从垃圾堆中救回了大量材料，同时也节省了使用其他材料的费用和精力。

INDEX

Ector Hoogstad Architects

Ector Hoogstad Architects (EHA) is an independent all round architectural practice based in Rotterdam, headed by Joost Ector (design principal) and Max Pape (managing director). EHA has a leading position in the Netherlands' internationally highly ranked design-landscape. In over fifty years, the firm has built up an impressive portfolio of projects, remarkable both for its range and variety as well as its consistently high level of both architectural and technical quality. The firm has a large number of striking and significant buildings to its credit, such as the Netherlands Ministry of VROM (Spatial Planning and the Environment), several theatres, and buildings for various universities.

SNØHETTA

Formed in 1989 and led by principals Craig Dykers and Kjetil Thorsen, Snøhetta is an award-winning international architecture, landscape architecture, and interior design firm based in Oslo, Norway, and New York City. As of 2010, the firm, which is named after one of Norway's highest mountain peaks, has approximately 100 staff members working on projects in Europe, Asia, the United States and Canada. Snøhetta is composed of designers and professionals from around the world. The firm has completed a number of critically acclaimed cultural projects, including the Bibliotheca Alexandrina in Egypt; the new National Opera and Ballet in Oslo, Norway; and the Lillehammer Art Museum, built for the 1994 Winter Olympics in Norway. Snøhetta was also recently commissioned to reconstruct the public spaces in and around New York City's Times Square, and has been selected to partner with the San Francisco Museum of Modern Art on the expansion of the acclaimed museum. In 2004 the company received the Aga Khan Award for Architecture, and in 2009 it was honored with the Mies van der Rohe Award. Snøhetta is the only company to have twice won the World Architecture Award for best cultural building, in 2002 for the Bibliotheca Alexandrina and in 2008 for the National Opera and Ballet in Oslo.

C. F. Møller Architects

C. F. Møller Architects is one of Scandinavia's oldest and largest architectural practices. Their work involves a wide range of expertise that covers programme analysis, town planning, master planning, all architectural services including landscape architecture, as well as the development and design of building components. Simplicity, clarity and unpretentiousness, the ideals that have guided their work since the practice was established in 1924, are continually reinterpreted to suit individual projects, always site-specific and based on international trends and regional characteristics. Over the years, they have won a large number of national and international competitions and awards. Their work has been exhibited locally as well as internationally at places like RIBA in London, the Venice Biennale, the Danish Architecture Centre and the Danish Cultural Institute in Beijing. Today C.F.Møller Architects has app. 320 employees. Their head office is in Aarhus and they have branches in Copenhagen, Aalborg, Oslo, Stockholm and London, as well as a limited company in Iceland.

Mecanoo Architecten

Mecanoo, officially founded in Delft in 1984, is made up of a highly multidisciplinary staff of over 90 creative professionals from 25 nationalities and includes architects, interior designers, urban planners, landscape architects and architectural technicians. The company is directed by its original founding architect, Prof. ir. Francine M.J. Houben and technical director Aart Fransen and are joined by partners Francesco Veenstra, Ellen van der Wal and Paul Ketelaars. The extensive experience gained over more than 25 years, together with structured planning processes results in designs that are realized with technical expertise and great attention to detail.

Francine Houben

Francine Houben, founder of the architects' firm Mecanoo Architecten in Delft, is one of Europe's most active architects. She lectures all over the world and takes part as a jury member in many prestigious competitions. She plays an important role in her designs innovation and sustainability. She combines architecture, urban planning and landscaping in an unorthodox way. Examples are the designs for the Philips Business Innovation Centre Fifty Two Degrees in Nijmegen, the Library Technical University of Delft, Municipal Offices and Station in Delft, the Wei-Wu-Ying Center for the Arts in Kaohsiung, Taiwan and the Library of Birmingham in the UK With her international staff of more then 90 Francine Houben is working on projects in the Netherlands, Spain, Taiwan, South Korea, Malaysia, UK, Poland, Albania, Italy, Japan, Denmark, Finland, Germany and Switzerland. Since 2000 Francine Houben is professor in Architecture and Aesthetics of Mobility at the Delft University of Technology. In 2007 she was visiting professor at Harvard University in Cambridge, U.S.A. She received numerous awards and recognitions such as honorary fellowships of the Royal Architectural Institute of Canada and the American Institute of Architects in 2007. In 2008 she was awarded the Veuve Clicquot Business Woman of the Year Award.

Hawkins\Brown

Hawkins\Brown is a London based architectural practice formed in 1988 by Roger Hawkins and Russell Brown. Since then they have grown steadily over the years to 15 partners and over 80 staff. They have a creative progressive award-winning studio that has a reputation as one of the leading design based architectural practices in the UK. Recently named AJ100 Practice of the Year 2011, Hawkins\Brown has never limited its architecture to a particular style type or scale of work. They bring a bespoke collaborative approach to their projects and a determinedly broad experience of building universities, schools, transport, housing, offices and arts projects across the UK and increasingly in Europe and further afield. In 2010 they published Ideas Exchange: The Collaborative Studio of Hawkins\Brown, presenting the recent work of the practice, communicating their philosophy that architecture benefits from the contributions of and collaboration with the wider team, including engineers, clients, artists, other architects and designers.Their work is regularly in the media; published in the design press, local and national papers and in international magazines. In addition Roger and Russell regularly contribute to and write for a number of architectural magazines and increasingly have been broadcast on television and radio.

AS.Architecture-Studio

Around its 12 partners, ARCHITECTURE STUDIO brings together about 200 architects, urban planners, designers, interior designers and economy planners of various nationalities. This collective dimension enhances the group's dynamic strength. In their opinion, architecture is primarily a social item. Architecture is the result of dialogue between economical, cultural and social interests which are interacting for each project. Dialogue also determines the philosophy of ARCHITECTURE STUDIO. This intellectual determination based on « leaving open the wounds of possibilities» (Kierkegaard) implies that before participating in any project, one must try to set free from all stylistic or technical presumption and remain open to the context's suggestions.
This step is set within the very heart of the process. This process is meant to be iterative and abstract, as well as organic and tangible.

Architecture-Studio

Architecture-Studio defines architecture as "an art committed with society, the construction of the surroundings of mankind". Its foundations lie on work group and shared knowledge, with the will to go beyond individuality for the benefit of dialogue and confrontation.
Thus, the addition of individual knowledge turns into wide creative potential. Founded in Paris in 1973, Architecture Studio brings together more than one hundred people, around 12 partners. This team of 25 different nationalities is composed of architects, urban planners, designers and interior designers. Architecture Studio has built itself upon the will of a will for broad-mindedness and group philosophy. Their team has remained and will remain open through time. The international presence of AS.Architecture-Studio is particularly strong in China. They have two permanent subsidiaries in Shanghai and Beijing. AS.Architecture-Studio China, which employs about 50 architects, carries out projects it has designed, while combining the mindset and skills of two cultures, French and Chinese. AS.Architecture-Studio works with large international groups such as Danone, Wison, Jinqiao Group, Solidere International, DAMAC Group, DCC…It develops big projects such as the City of Science in Chongqing, as well as Souria Towers in Damascus. For several years now, AS.Architecture-Studio has been developing its interior design department, AS Design Studio, which prolongates AS's architectural concepts, as well as the design of prestigious projects such as the renovation of a town house located Place Vendôme in Paris, the renovation of Pallazzo Santa Maria Nova in Venice, the design of Bali Barret shop in Tokyo and the development of luxury villas in Dubai.

mdu architetti

mdu architetti is an associated firm established in Prato, Italy in 2001 by Valerio Barberis (1971), Alessandro Corradini (1964) and Marcello Marchesini (1970), later joined by Cristiano Cosi (1974). As regards their design approach, the members of mdu say: "The sphere of activities includes contemporary landscapes, physical and non-physical places now inhabited by humans. The method is based on doubt, the never-ending discussion of all that appears obvious, usual and certain, in the conviction that there are many invisible cities besides the official one. The approach is aimed at seeking new trajectories through a different, transgressive look at each landscape. The purpose is to pierce reality: architecture is a collision between the identities that comprise each landscape." mdu's works have received many awards and much recognition, and have featured in national and international publications. Completed works include: the Poolhouse Fioravanti in Prato, the RRS & Feng Lin showroom in Shanghai, the EsseBi showroom in Agliana (PT), the Contemporary Art Gallery of Florence dedicated to Giuliano Vangi, the transformation of industrial buildings into lofts in Prato, the Municipal Library of Greve in Chianti (FI), and the Theatre of Montalto di Castro (VT). The members of mdu are also dedicated to research and teaching. Valerio Barberis and Marcello Marchesini (who both have PhDs in architectural design) are currently adjunct professors in Architecture Design at the Faculties of Architecture of Florence and Parma. Alessandro Corradini holds a fellowship in art and architecture at the Schloss Solitude Academy in Stuttgart.

Keith Williams

Keith Williams is the award-winning founder and design director of London based Keith Williams Architects. He is recipient of more than 25 major design awards and has twice been BD Public Building Architect of the Year in both 2006 & 2008. He works internationally on major civic, arts and masterplanning projects.
His recent projects include the new Marlowe Theatre, Canterbury, and the new Chichester Museum, whilst his many celebrated completed projects include Athlone Civic Centre and Wexford Opera House both in Ireland, and the Unicorn Theatre and the Long House, both in London. Under construction inter alia his project for Athlone Art Gallery (Ireland) will finish later this year. Most recently has prepared designs for a new city for 10 million inhabitants near Karachi in Pakistan. Keith Williams has lectured widely on his work and in 2010 was made Honorary Visiting Professor of Architecture at Zhengzhou University, China. He is a Fellow of the Royal Society of Arts, a Member of the Royal Institute of the Architects of Ireland, a member of the National Review Panel at Design Council CABE and he sits on the National Panel of the Civic Trust Awards. He has judged numerous architectural competitions and awards schemes and his work has been published worldwide. The first monograph on the firm's work entitled "Keith Williams: Architecture of the Specific" was released in December 2009 by Images Publishing Ltd, Melbourne, Australia.

ARQUITECTONICA

Arquitectonica is a full-service architecture, interior design and planning firm that began in Miami in 1977 as an experimental studio. Led by Bernardo Fort-Brescia, FAIA, and Laurinda Spear, FAIA, ASLA, the studio has evolved into a worldwide practice, combining the creative spirit of the principals with the efficiency of delivery and reliability of a major architectural firm. Its affiliated firm, ArquitectonicaGEO, provides landscape architecture services. Today Arquitectonica has a practice across the United States directed from regional offices in Miami, New York and Los Angeles. Arquitectonica's international practice is supported by their European regional office in Paris; Asian regional offices in Hong Kong, Shanghai, Manila; Latin American regional offices in Lima and São Paulo; and the Middle East office in Dubai.

Ben van Berkel / UNStudio

Ben van Berkel studied architecture at the Rietveld Academy in Amsterdam and at the Architectural Association in London, receiving the AA Diploma with Honours in 1987. In 1988 he and Caroline Bos set up an architectural practice in Amsterdam. Ben Van Berkel & Bos Architectuurbureau has realized amongst others projects the Karbouw office building, the Erasmus bridge in Rotterdam, museum Het Valkhof in Nijmegen, the Moebius house and the NMR facilities for the University of Utrecht. In 1998 Ben van Berkel and Caroline Bos established a new firm: UNStudio. UNStudio presents itself as a network of specialists in architecture, urban development and infrastructure. Current projects are the restructuring of the station area of Arnhem, the mixed-use Raffles City in Hangzhou, a masterplan for Basauri, a dance theatre for St. Petersburg and the design and restructuring of the Harbor Ponte Parodi in Genoa. With UNStudio he realized amongst others the Mercedes-Benz Museum in Stuttgart, a façade and interior renovation for the Galleria Department store in Seoul and a private villa up-state New York. Ben van Berkel has lectured and taught at many architectural schools around the world. Currently he is Professor Conceptual Design at the Staedelschule in Frankfurt am Main and for the Spring term 2011 was awarded the Kenzo Tange Visiting Professor's Chair at Harvard University Graduate School of Design. Central to his teaching is the inclusive approach of architectural works integrating virtual and material organization and engineering constructions.

M+R interior architecture

M+R interior architecture is an international operating office founded in 2000 by Hans Maréchal. Their fields of activity often involve complex assignments such as converting and designing offices, airports, libraries, restaurants, hotels, theatres and shops. Among their design skills and core activities for building and interior architecture they are also involved with revitalizing existing buildings and monuments in particular. The architects from M+R determine the form and content of each design assignment on the basis of the programme of requirements. Creativity, functionality, sustainability and ergonomics are translated in a well thought-out manner into a unique final product with an identity of its own. The starting-point is usually a conceptual approach to the design task, for which a total plan is developed, with a sharp eye for details. The power of a strong design is vision, innovation and the quality of realization. At an early stage they are able to provide their clients with a good picture of the final result by means of their computer visualizations. They attach a lot of value to the integration of technology and its possibilities. Light architecture is also a speciality within the firm, seeing as light in particular makes an important contribution to the quality of life.

DONAIRE ARQUITECTOS

DONAIRE ARQUITECTOS was born in Seville in 2000. Youth team with a large experience working in innovated architecture with traditional and new materials. They design public and private buildings, furniture and graphic design.

Paul Andreu

He has designed and built many airports. He has worked without interruption, over a period of thirty years, on the original conception and later development of Paris' Charles-de-Gaulle airport. Going back over the same subject again and again, becoming "a specialist", is good schooling in what it means to be serious and humble. This constant returning to the self-same demanding problem clarifies much about what architecture is, about the necessity for it to be grounded in usage and functionality as well as in technique and construction, and the even greater necessity for it to surpass them all in order to exist in the realm of intelligence and art. What he is seeking in any project is at once its inner coherence, its intelligibility, and its relationship to the outside. He regards each project as a complete, self-enclosed world and, at the same time, as but a part of a vaster whole that can be linked to the physical place, the site, and more generally to the environment, but often as well to a whole that only the mind is capable of reconstructing on the basis of scattered elements. Another idea which is a prime mover in his work is that when a thriving architectural structure leaves the hands of the architect, it is in unfinished state. To bring it to completion, it must be confided to the elements: to light, to wind, to water. His work with light in the T.G.V. (high-speed train) station at Charles-de-Gaulle airport and in the upcoming extension to Terminal 2 is to be regarded from this standpoint. He is interested in attaining a sense of weightlessness and transparency and he strives to tackle all the details of construction with great precision and truthfulness. But to him what matters most in the space itself, its structure and its bounds as defined by the material which stands out against the light or dissolves into it.

studio101 architects

studio101 architects is a Geelong based, multi award winning, innovative and progressive architectural practice led by founding director Peter Woolard. studio101 architects pursues a contemporary and refined form of sustainable architecture, based on a modernist approach to composition and detail. The practice offers a thorough and professional scope of services from Concept Design through to Contract Administration during Construction. Their wide ranging experience includes Commercial, Residential and Interior projects located in a mix of urban, rural and coastal environments across Australia.

Diamond Schmitt Architects

Established in 1975, Diamond Schmitt Architects Incorporated has achieved national and international recognition for innovative design excellence in a broad range of building types. The firm is known and respected for responsible project management and budget control. Since 2004, they have been the only Canadian architecture firm to be named among Canada's 50 Best Managed Companies. A dedicated and principled approach to all aspects of design delivers buildings that consistently meet and exceed client expectations. Diamond Schmitt Architects is a leading Canadian architectural practice recognized for excellence in the design of award-winning performing arts centres, academic and research buildings, commercial, residential and health care institutions. Current projects include the New Mariinsky Theatre in St. Petersburg, Russia, Sick Children's Hospital Research Tower in Toronto and the St. Catharines Centre for the Performing Arts.

V+

Towards Increased Well-being (VERS PLUS DE BIEN êTRE / V+) studio was founded in Brussels in 1999. Published both in Belgium and abroad, exhibitions include Portugal, Germany, France, Italy, and China. V+ had a presence in the associative and cultural milieu, supporting artistic projects or citizens projects and temporary applications. It carries out new projects and renovations, devoting considerable attention to spatial and technical experimentation. In the public domain, especially in the case of competitions, it undertakes the realisation of projects with the major symbolic, cultural or social importance. Architecture is not a question of square meters or provision of services. It is first and foremost a political stance, a cultural act, a poem of centimeters, a social statement, a philosophical surprise, a desire for space, a source of dreams, and above all about more life, more passion, more attitude, more joy, more questions, more intensity, more assertions, more fantasies, more euphoria, more effervescence, more movement, more resistance, more instinct, more desires, more character, more will, more demands, more eagerness, more spirit, more sparkle, more freshness, more fulfillment, more vivacity, more conviction, more pride, more generosity, more love, more temerity, more audacity, more delight, more playfulness, more infatuation, more challenge, more sensuality, more fascination and more inspiration.

ARRIS ARCHITECTS

ARRIS Architects was setup in January 2004 By Directors Shubhashish Modi & Satish Shetty. While targeting design capabilities of every individual architect in the firm, Arris has conveyed a gamut of designed buildings and spaces throughout India. These entities include Educational Institutions, Malls, and Multiplexes, Amusement parks, Residential Complex, Corporate Offices and more. Shubhashish teams up with the clients to get the initial project requirements, understanding their business and preparing design parameters. His work has been instrumental in conceptualization of various architecture and interior projects built by the firm. The building takes form on Satish's workdesk. He ensures that Form is in sync with Facts and Functions. With his sound knowledge of latest building technology and construction, Arris has been instrumental in completing around 150 Screens across India. As a team, Arris is a collaboration of individuals with excelling capacities in various facets of Architectural, Urban & Design practices. This design-orientated practice has built a reputation for High Quality Design Solutions delivered within record time schedules. Distinctive aspect of the firm has been an accurate interpretation of the client's requirements and converting them into reality through cost effective design solutions. ARRIS Architects' philosophy is built on the dual concepts of client service and design excellence. Aspect of energy efficiency forms an integral part of the design process and is reflected in the company's commitment to incorporate advanced simulation tools to predict the performance of the proposal at the conceptual design stages.

Robert Majkut

Robert Majkut (born in 1971) – designer, the founder and CEO of Robert Majkut Design studio – one of the most acknowledged creative company in Poland, an expert in designing for business. The quality of design solutions, unique methods of working and the understanding of the strategic importance of design to business made Robert Majkut become a partner for the most demanding investors. The creative team under his guidance works for the elite of business, both in Poland and abroad, in China and Russia. As a perfectionist, Robert Majkut rebels against mediocrity and ugliness; without any compromises seeks the highest quality in design. In his work he is not afraid to act out of conventions and often introduces progressive and innovative solutions. For Robert Majkut design is something more than just designing. It is a great responsibility for shaping the present and future. It is an aspiration to change the world for the better. It is a task of thinking about the human being and the environment in ways that meet the challenges of tomorrow.

Eranna Yekbote

B.Arch. University Of Mumbai, India 1998.COA. As the founding principal, Eranna provides leadership, guidance to all building design efforts at Era Architects. He is involved in conceptual design through completion, of every project. Eranna brings over 10 years of experience on a broad range of project types and sizes. After graduating from Mumbai University in 1998, he began his Architectural career with T.Khareghat and Associates, working on residential projects and Multiplexes in Mumbai. He started Era Architects in 1999 and has since then developed a reputation for creative design solutions for complex urban projects ranging from Multiplexes, Shopping malls, Hotels and residential projects. He is also responsible for marketing the design capabilities of the firm, advancing the corporate initiatives, while overseeing the operational and quality objectives office. The office is located in Central Mumbai and has a staff of twenty. In-house capabilities include master planning, urban design, and full-service architectural interior design space planning and graphic design. The firm maintains hands-on involvement with each project through all phases. This commitment has lead to the firm's success with repeat clients. Era Architects has completed projects in India, China, Philippines, Oman, Bangladesh and Kenya.

James Law

James Law is the visionary founder and Chief Cybertect at James Law Cybertecture International, formed on the first day of the new century in 2001. Dedicated to blend cyber technology within modern architecture, James is paving the way for people to live the future in a seamless and enjoyable manner. Graduated with a degree in architecture from University College London in 1992, James consolidated his knowledge in architecture, buildings, design, and technology at prestigious Japanese architect firm Itsuko Hasegawa in 1994 and became Director at world renowned design firm Gensler International in 1997. James' prime interest lies in shaping the future direction of people's everyday life. Consequently, he developed a core concept of Cybertecture, which is inspired by a fusion of advance technology and futuristic aspirations in buildings and products for the next generation. In reward for James's passion and contribution, James has won numerous international awards on architectural design and leadership, highlighted by the achievements of Young Global Leaders 2010 in World Economic Forum and CNBC International Architecture Awards during 2009. Within the architecture spectrum, he achieved the Royal Institute of Architects Award in 2003, which also earned him to become a member of the Royal Institute of British Architects. James achievements are further highlighted with Asian Innovation Awards 2004, CNBC International Property Awards 2007, and CNBC Asia Pacific Commercial Property Awards 2009.

João Luís Carrilho da Graça

João luís carrilho da graça, architect, graduated from esbal (lisbon higher-education school of fine arts) in 1977, the year he initiated his professional activity. Leactured at the faculty of architecture of the technical university of lisbon between 1977 and 1992. professor at the autonomous university of lisbon from 2001 to 2010 and at the university of évora since 2005. headed the architecture department at both institutions until 2010. Invited professor at the higher technical school of architecture of the university of navarra in 2007 and 2010. Invited lecture in seminars and conferences at several international universities and institutions.

Awards :

The title of "chevalier des arts et des lettres" by the french republic in 2010;

The "pessoa award" in 2008,

The "luzboa-schréder" award in 2004 at lisbon's first international light art biennal, the order for merit of the portuguese republic in 1999.

The "international art critics association" award in 1992.

Marcy Wong and Donn Logan

Marcy Wong received her B.A. in Art History from Columbia University, Barnard College in New York City. She obtained her graduate (M.Arch.) degree in architecture from the Columbia University Graduate School of Architecture. Ms. Wong founded her own firm in 1986, and in 1999 Donn Logan joined the practice to form the current partnership in Berkeley, CA. Their work together covers a diverse array of building types, including many projects for cultural and arts facilities.

Donn Logan launched his professional career soon after graduation from Harvard University's Graduate School of Design, with commissions resulting from design competitions. These projects led to the founding of his first firm, ELS / Elbasani, Logan & Severin. Mr. Logan led that firm to be named California firm of the year by the California Council AIA in 1991. While a principal at ELS, Mr. Logan was lead designer for the firm's civic, educational, and recreational design work. Over a decade ago, he left that firm to form his current partnership with Marcy Wong. Throughout his career, Mr. Logan has been responsible for many projects which have received design awards, including seven prizes in major design competitions, Progressive Architecture citations, numerous AIA awards at the national, state and local levels and five awards from the USITT.

HENNING **LARSEN** ARCHITECTS

Henning Larsen Architects is an international architecture company with strong Scandinavian roots. Their goal is to create vibrant, sustainable buildings that reach beyond themselves and become of durable value to the user and to the society and culture that they are built into. Since the company was founded in 1959, they have acquired a comprehensive knowledge of the many aspects of building – from sketch proposals to detailed design, building owner consulting and construction management. They shape, challenge and change the physical environment – from masterplans, urban spaces and buildings to interior design, components and strategic design – with the overall objective of providing the user with a strong, visionary and thoroughly prepared design adapted to the specific context. Henning Larsen Architects attaches great importance to designing environmentally friendly and integrated, energy-efficient solutions. Their projects are characterised by a high degree of social responsibility – not only in relation to materials and production but also as regards good, social and community-creating spaces. Their ideas are developed in close collaboration with the client, users and partners in order to achieve long-lasting buildings and reduced life-cycle costs. This value-based approach is the key to their designs of numerous building projects around the world – from complex masterplans to successful architectural landmarks.

Niels Van Eijk & Miriam Van der Lubbe

"Our designs raise questions; we work with forms and products we all know, we add benefits to them to make it nowadays products without loosing their own characteristics. They make you change your perspective to things, and challenge you to look at another way to the world around you." Graduated at the Design Academy and the post graduate programme of the Sandberg Institute in Amsterdam. Started their design studio in 1998. They work individually but share one studio. Often they collaborate in projects. They focus on product, interior and exhibition design. Work has been exhibited worldwide and purchased by many museums such as Museum Boymans van Beuningen Rotterdam, Centraal Museum Utrecht, Museum voor Moderne Kunst Arnhem, The Dutch Textile Museum Tilburg, Manchester City Art Gallery, WOCEF Korea, Museum FIT New York, Princessehof Leeuwarden and Stedelijk Museum Amsterdam.

Arthur Chan

Arthur studied architecture in University of Hong Kong and received his Master of Environmental and Sustainable Design in Chinese University of Hong Kong and Master of Fine Arts in RMIT University. In 1995, he found DPWT design Ltd in Hong Kong with his partner and subsequently established branch offices in Beijing and Shanghai in 1999 and 2003. He is currently the lecturer/program coordinator of the Hong Kong Art School. His artworks had been collected privately. His works is widely published in London, Italy, Australia, and throughout Asia in Singapore, Malaysia, mainland China and Hong Kong. He was invited to feature his works in 1000 architects in Australia. Arthur set up his own practice in 1995 and has offices in Beijing, Shanghai and Hong Kong. He is Doctor of Fine Arts candidate in RMIT University. He works and lives in Beijing/Hong Kong. Arthur has earned a string of competition successes, including the recipient of:

2011
- Jintang Prize – Top 10 Public Space Design of the Year - GH Citywalk Cinema
- Jintang Prize – Good Design of the Year, Public – GH Citywalk Cinema
- Successful Design Award – GH Whampao Cinema
- 3rd Lighting Design Competition – GH TsingYi Multiplex

2010
- Nest Award in 2010 International Space Environmental Art Design Competition – GH Tsingyi Cinema
- Asia Pacific Interior Design Biennial Awards – GH Tsingyi Cinema
- IF Design Awards China 2010 – GH TsingYi Cinema
- Nest Award in 2010 International Space Environmental Art Design Competition – GH Citywalk Cinema
- The Most Successful Design Award – GH Citywalk Cinema

2009
- The Most Successful Design Award – GH TsingYi Cinema
- APIDA – top 10 of Commercial Category – GH TsingYi Cinema

Glyn Mellor

Glyn, a director at NBDA Ltd since 1994, has nearly 30 years of experience working on cinema and leisure projects. He has carried out projects throughout the United Kingdom and Europe. He has worked with a variety of clients and operators ranging from the biggest chains, such as Odeon Cinemas Ltd, to smaller independent companies such as WTW Cinemas, Cine-Bowl, Firoka Leisure and Savoy Cinemas. His work on multiplex cinema projects often involves coordinating the design with the developers' overall proposals, often within large mixed use developments. Existing cinema subdivision and refurbishment requires a detailed knowledge of historic buildings and current cinema requirements to adapt the existing building in the most effective manner. This experience has led him to work on the most prestigious projects such as the refurbishment of the Odeon Leicester Square and Mezzanine.

ACKNOWLEDGEMENTS

We would like to thank everyone involved in the production of this book, especially all the artists, designers, architects and photographers for their kind permission to publish their works. We are also very grateful to many other people whose names do not appear on the credits but who provided assistance and support. We highly appreciate the contribution of images, ideas, and concepts and thank them for allowing their creativity to be shared with readers around the world.

后记

本书的编写离不开各位设计师和摄影师的帮助，正是有了他们专业而负责的工作态度，才有了本书的顺利出版。参与本书的编写人员有：

ADPi, AETER, Ajinkya Manohar, Alain Janssens, Alessandro Corradini, Andrey Cordelianu, Architecture Studio, ARQUITECTONICA, Arris Architects Pvt Ltd, Arthur Chan, AS. Architecture-Studio, Batteriid Architects, Ben van Berkel, Bevin Chen, Billy Hustace, Birdseyepix, C. F. Mřller Architects, Carmela Domínguez Asencio, Celine Nelke, Chester Ong, Christian Richters, Cristiano Cosi, Curtis & Rogers Design Studio, David Rapado Moreno, Delia Pacheco Donaire, Diamond Chan, Diamond Schmitt Architects, DL Structures , DPWT Design Ltd., ECADI, Ector Hoogstad Architecten, ERA Architects, Fernando Alda, FG+SG Fotografia de Arquitectura, Frank Tielemans, G Mellor, Glyn Mellor, Hack Kampmann, Hawkins\Brown, Helene Binet, Henning Larsen Architects, Hufton+Crow, Ignacio Núñez Bootello, Iwan Baan, Jack Diamond, James Law, Jan Hoogstad, Jeroen Musch, Jestico+Whiles, Jesús Núñez Bootello, João Gomes da Silva, João Luís Carrilho da Graça, Juan Pedro Donaire Barbero, Julian Weyer, Kavi Sanchez, Keith Williams Architects, Lorenzo Boddi, LUC BOEGLY, M+R interior architecture, Marcello Marchesini, Marcy Wong Donn Logan Architects, mdu architetti, Mecanoo architecten, Mr. Eranna Yekbote, Mr. Girish Patil, Mr.Vikas Jain, Mumbai, NBDA Architects , Nic Lehoux, Nuno Gusmão, Pablo Baruc García Gómez, Paul Andreu Architecte, Paul French, Pietro Savorelli, Prague, Robert Majkut, Robin Hill, ROR Studio, Shai Gil, Sharon Risedorph, Snřhetta AS, studio101 architects, Szymon Polański, The Hall Group, Tibisay Cañas Fuentes, Tim Crocker, Tim Stubbins, Trevor Mein, UNStudio, V+, Valentina Muscedra, Valerio Barberis, Van der Lubbe, Van Eijk, Willie Wu